Advance Praise for College Media

"A great injustice is finally righted with *College Media: Learning in Action*, as Gregory Adamo and Allan DiBiase powerfully shine the spotlight on the often overlooked impact of the college media experience.

As evidenced in the wide-ranging collection of essays presented here, many generations of Americans have genuinely benefited from once participating in college media organizations. Whether it was by working on a student newspaper, radio or television station, participating in college media was transformative in both their professional and personal development. *College Media: Learning in Action* is effective in showcasing just how important student media environments are in helping students develop their own skills and talents for their future careers.

College Media: Learning in Action also provides valuable insight into the development and continuing influence of college media upon campus communities and society as a whole. Through offering both historical and theoretical perspectives, Adamo and DiBiase provide the best explanation yet of how college media organizations have historically developed and survived tumultuous changes to become the places today where passionate students still find their own voices and form their self-identities.

College Media: Learning in Action offers us fresh perspectives with which to evaluate the influence that college media continues to have on the current generation of students, faculty, and staff in the country, and is a truly welcome and substantive effort towards finally giving it its due recognition."

—Rob Quicke, Associate Professor of Communication,
William Paterson University; Founder, College Radio Day

"It's been said that advising college media is the loneliest job on campus. Consider this guide an essential companion for anyone seeking to navigate challenges and changes in a complex, evolving environment. From Dewey to digitization to diversity training, this resource unpacks the varied learning experiences in our college newsrooms and radio stations that will help prepare future leaders and communicators in 'laboratories of innovation and courage.'

The triangulated approach affirms the value of student media as a model for collegiate-level experiential learning that instills lifelong transferrable skills, guided by theoretical underpinnings from academic and practitioner lenses. The essays offer a personal glimpse into these environments that mold and shape future media professionals, providing practical, nuts and bolts insight and ideas that will strengthen campus communities through the breadth of issues in the college media landscape today.

This anthology is a veritable 'Who's Who' of college media scholarship and influential thought, compiled into one comprehensive resource, that will be used to inspire, affirm, and improve communication practices throughout campuses nationwide."

—Lisa Lyon Payne, Associate Professor and Coordinator of Communication
and Adviser to The Marlin Chronicle, *Virginia Wesleyan College;*
Research Chair, College Media Association

College Media

This book is part of the Peter Lang Media and Communication list.
Every volume is peer reviewed and meets
the highest quality standards for content and production.

PETER LANG
New York • Bern • Frankfurt • Berlin
Brussels • Vienna • Oxford • Warsaw

College Media

Learning in Action

Edited by Gregory Adamo and Allan DiBiase

PETER LANG
New York • Bern • Frankfurt • Berlin
Brussels • Vienna • Oxford • Warsaw

Library of Congress Cataloging-in-Publication Data
Names: Adamo, Gregory, editor of compilation. | DiBiase, Allan, editor of compilation.
Title: College media: learning in action / edited by Gregory Adamo, Allan DiBiase.
Description: New York: Peter Lang, 2017.
Includes bibliographical references and index.
Identifiers: LCCN 2016032506 | ISBN 978-1-4331-2431-0 (hardcover: alk. paper)
ISBN 978-1-4539-1831-9 (ebook pdf) | ISBN 978-1-4331-3616-0 (epub)
ISBN 978-1-4331-3617-7 (mobi)
Subjects: LCSH: College student newspapers and periodicals—United States.
College radio stations—United States. | Journalism—Study and Teaching—United States.
Classification: LCC LB3621.65 .C65 2017 | DDC 378.1/9897—dc23
LC record available at https://lccn.loc.gov/2016032506
DOI: 10.3726/9781453918319

Bibliographic information published by **Die Deutsche Nationalbibliothek**.
Die Deutsche Nationalbibliothek lists this publication in the "Deutsche Nationalbibliografie"; detailed bibliographic data are available on the Internet at http://dnb.d-nb.de/.

Table of Contents

Introduction

GREGORY ADAMO AND ALLAN DIBIASE

College newspapers, radio stations, and television outlets operate as unique spaces that allow students learning and leadership opportunities unobtainable elsewhere on most campuses. Students become exposed to issues as varied as planning, group dynamics, meeting deadlines, and coping with regulatory structures as well as larger matters like the benefits of diversity. Many of these features are ones students will encounter should they go on to careers in the media, but college media work also prepares individuals for almost any work situation that involves small group interaction, supervisory situations, and dealing with institutional pressures and constraints.

In this volume we are interested primarily in detailing the variety of ways students learn through participation in such organizations and making this available as justification for support of these rich, alternative learning opportunities. As colleges and universities assess their commitment to this kind of learning, and as newer forms of social media become more prominent, it becomes increasingly important to understand and describe what happens in these unique spaces lest they become assimilated into more ubiquitous templates for learning, or, eliminated completely as we next discuss.

In part, then, the point of departure in collecting the essays in this book was located in the evolving media landscape that has altered many traditional features of the undergraduate experience in media production. College newspapers are scaling back or eliminating print editions in order to go online, and FM licenses for college radio stations have been sold off or the stations shut down. Overall, the media consumption habits of both students and the public at large are changing rapidly as they have become accustomed to the immediacy of news and information via smartphones and tablets.

Threats to the independence or even the existence of college media entities have received attention over the past several years. In 2012 much of the staff of the University of Georgia's student newspaper, *The Red and Black*, including the top editors, resigned, claiming "interference, even censorship, by the nonstudent managers hired to oversee it" (Pérez-Peña, 2012, para. 1). The student journalists felt they were being pressured to be more of a public relations vehicle for the University.

College radio has a long and extensive history of getting squeezed out by more powerful forces. Beginning in the early 1970s National Public Radio built its foundation on professionalizing student-operated educational FM stations and lobbying the FCC to change its rules on low-power FM, thereby forcing many college stations off the air. In 1978 the FCC issued an order that required 10-watt stations to increase power to 100 watts or subject themselves to encroachment from higher-power stations. This eliminated a relatively inexpensive way for colleges to broadcast. It was also a step toward professionalizing college radio, often eliminating any student involvement. According to Dunbar-Hester, the Corporation for Public Broadcasting and NPR "sought to consolidate the FM band to expand NPR in large part through absorbing existing stations into the network" (2014, p. 6).

Since 2010 the University of San Francisco, Rice University, Georgia State, and Vanderbilt University all sold their FM radio stations. *The Chronicle of Higher Education* identified this as a trend that started in the 1990s when a number of college stations, "were driven onto the Web or into oblivion when college administrators decided to sell their licenses for much-needed cash" (Troop, 2011, para. 2). Vanderbilt used the $3.35 million from the sale of WRVU to finance an endowment for student media at the university, including publications and a version of WRVU that would be available to those with HD radio receivers. The endowment was seen as necessary since other funding sources, such as advertising in the student newspaper, were drying up.

These developments have become obvious in all forms of student media across the nation. In 2012, as the *Washington Times* pointed out,

> College papers at schools as diverse as Bowdoin College in Maine, the University of Connecticut and the University of Texas face a battle for shrinking student activity dollars in an age of reduced state funding. Students at the University of Nebraska-Omaha this week voted to retain funding for the twice-weekly *Gateway* after the staff warned that a cut could mean the death of the 99-year-old paper edition. (Jackson, 2012, para. 9)

The ever tightening financial situation for many colleges may be the impetus for selling off some of these resources, but these actions have also caused negative publicity for the institutions, and bitter feelings among alums, some vowing not to donate to their alma mater. And then there is the core issue that this volume addresses: The loss of valuable creative and learning spaces for students (Student Press Law Center, 2014).

Absent in much of the discussion of these developments is consideration of student media as an important resource in the educational mission of the university. As essays in this volume will show, college newspapers, news bureaus, television operations, and broadcast stations function as learning laboratories, places where students learn leadership and team building skills and competencies essential to a wide variety of careers. Many of us believe the availability of these resources on campuses should continue to play a role in students' education. For example, in 2015, the journal *Interactions: Studies in Communication and Culture* devoted an entire issue to college radio, claiming in an editorial that "Making radio can make students feel more confident, boost their self-esteem and give them the possibility to speak about their cultures, their social groups and their lives, and can help to provide a local view on global issues" (Scifo, 2015, p. 5).

One reason for this book is a need to explore the role of college media in a student's education and to bring these findings to those making decisions about these entities. As editors, we also wanted to present this argument to those more directly involved, be they faculty advisors, student life professionals, or the students themselves. It is not until many years after their college media work that many alums realize how important their involvement was to their education. Making this clear to students at the time of their involvement might also be helpful to those advisors struggling to help students see beyond writing their latest news story, producing a video package, or bringing the new release from a hot band to the airwaves. We decided to solicit essays that provide the educational underpinning—not just for traditional broadcast and print media organizations but also for those online—addressing how students learn through voluntary participation in undergraduate media organizations even as these media are changing and evolving.

Anyone associated with professional media knows that many people in these fields got their first experience and often, career orientation, through participation in their college radio station, television operation, newspaper, or news bureau. What is it about these undergraduate experiences that have the power to generate careers? The research, stories, and case studies in this book are designed to explore this educational resource found on college campuses throughout the country.

As a response to the issues mentioned above, the authors contributing to this volume use a multifaceted lens to reveal how people learn through experiences that do not happen in the classroom and are not packaged as "educational." The current emphasis on formulation, delivery and testing of formal educational experiences has perhaps obscured the fact that learning happens, often most effectively, in other ways and places. Although the focus of this volume may seem obvious and remedial, it gets its enduring relevance from a famous quip made by Hegel (1977, p. 18) in his *Phenomenology of the Spirit*: "Quite generally, the familiar, just because it is familiar, is not cognitively understood." So it often is with the notion of learning through experiences when not pre-identified as "education."

This volume is divided into four sections: Theory; Description and Narrative; Case Studies; and Contextual Issues. These sections provide foundational understandings, first-person accounts, discussions of particular media organizations, and, finally, issues of concern that affect how students learn through their experiences in media organizations. Following are brief descriptions of these sections and the essays in each.

THEORY

The volume begins with a foundational essay on learning through experience based on the work of the American philosopher and educator John Dewey. Although Dewey's work is familiar to generations of educational theorists, his particular emphasis on a fuller understanding of how people have experiences in particular situations and contexts, such as the ones discussed in this book, is enduring and important. The author, Allan DiBiase, elaborates on Dewey's work by using a pluralistic and diverse theoretical model of "what experience is" to highlight the unique and situational aspects that make up learning experiences. It is with this holistic model in mind that the learning experiences examined in this book can best be understood.

Melony Schemberger's essay, "College Media as an 'Extracurricular Activity': Historical and Theoretical Perspectives" illustrates how theorists like Alexander Astin and others, building on the generative work of Dewey, have theorized the learning opportunities in extracurricular involvement at the undergraduate level of study. This essay makes specific both the rationales and benefits of college learning experiences gained outside the classroom.

Opportunities to practice and learn leadership skills while participating in college media organizations are understudied and often unrecognized except anecdotally. "College Media and Leadership Training: Conceptualizing Student Leadership" by Laura K. Dorsey-Elson theorizes and demonstrates how this highly valued form of learning is made available through participation in college media organizations.

DESCRIPTION AND NARRATIVE

The three essays in this section reflect the significance of ethnography as a means of capturing the fullness of human experience. This is achieved in a variety of ways including the authors' personal narratives and their interviews with student participants about their experiences. The essays provide detailed accounts that expand and supplement what is often conveyed in anecdotal shorthand regarding the value of college media learning experiences. These essays provide views that are inclusive and qualitative about the processes of learning through and in experience.

The first essay in this section, by Len O'Kelly, opens with a direct personal statement reflecting an observation made in this Introduction about how individuals get their career starts in college media organizations. The essay, "Why College Radio," begins in O'Kelly's own experience and career but the focus is rapidly widened—applicable to other forms of media—by interviews with alumni and current students that emphasize particular skills and competencies gained through their involvement in college media organizations.

Leo Fahey's essay provides a richly detailed personal account of his experiences in "early college radio." The essay then examines, over time, through use of the work of Irving Goffman, aspects of learning that are rarely discussed in the literature. In this way the essay delivers on critical elements of the model presented in the opening essay in this volume: The emotional, psychological and sociological factors that affect how and what students learn.

Leigh Landini Wright's "Lessons Learned: How College Newsrooms Prepare Students for the Professional Newsroom and Beyond" is the first of two essays (see Chapter 15) that examine in part the "hospital teaching model" of journalism education; finding that "learn by doing" is a common focus in the model and in traditional views of how students learn in media organizations. Through extensive interviews with alumni Landini highlights the fundamental skills and competencies learned through college media involvement and their broad applicability to careers other than communications and media.

CASE STUDIES

The five essays in this section offer case studies to the different ways that students learn. In so doing, they reveal the diversity of ways in which media organizations exist and function, as well as the learning opportunities that follow from this involvement.

"Student Newsmakers: A Case Study in Socialization and Professionalization" uses a qualitative interpretive approach to better understand learning in student media organizations by going inside one as an ethnographic observer. Kimberly Meltzer's descriptions shed light on how, through processes of socialization, students learn skills and dispositions that are considered essential to professional media careers. This essay provides a rare, detailed *in situ* look at a college newsroom and the dynamics that affect learning.

In "Laws, Rules and Regulations: Student Media Learning Through Obstacles, Obstructionism and Opacity," Susan Kirkman Zake discusses cases involving misuse of FERPA regulations, barriers to accessing public records, and many increasingly hostile administrative responses to student press coverage of significant events on college campuses. These cases provide a vivid picture of both the obstacles and cultivation of the dispositional grit required to function effectively as a student journalist today. Zake's account is a clear example of how student media can provide real world experience for those looking to join the journalism profession.

Haijing Tu's "The Learning Curve at Student Media: Old Challenges and New" examines student media organizations at Indiana State University showing how the experiential learning associated with traditional media organizations—characterized by both the freedom to experiment and learning skills essential to professional development—are shifting under the forces of new social media entities. This essay also examines the relationship of student media organizations to journalism and communication curricula.

In "Dealing with Disruption: Learning Industry Lessons on the Job in Student Media," Susan Kirkman Zake presents sharply contrasting case studies in dealing with "innovative disruption" at the University of Oregon's *Oregon Daily Emerald* and the Western Kentucky University's *College Heights Herald*. These case studies focus on how and what students learn as traditional media outlets cope and change with new and innovative modes of publication.

In "'*Welcome to The Drew Vault*': Educating Students through College Radio" Christopher J. Anderson demonstrates the use of radio as an enriching educational tool in a variety of innovative ways that knit together various elements of a college campus. This case study provides an example of how student media organizations can find relevance, new audiences and unique learning opportunities—for participants as well as audiences—by exploring themes and resources considered essential to a liberal education.

"The Shock of the New: A Case Study of Technological Innovation at a Student-Run University Television Station," by Megan Hopper and John Huxford, examines the uses of technology at Illinois State University in shaping both the opportunities and "headaches" of how student engage with and learn from a rapidly converging media environment. The essay is focused through discussion about the implementation, function, and learning available through a "Social Media Analytics Command Center (SMACC) which is a state-of-the-art computer lab used to follow and analyze public conversations from social media outlets—including Facebook, Twitter, Instagram, Reddit, YouTube, and Google Plus—as well as monitoring popular discussion forums and blogs.

CONTEXTUAL ISSUES

The final section of this volume discusses topics and issues that affect college media organizations and subsequently the quality of learning opportunities available through them. The topics involve how media organizations are advised, how they can be shaped by better understanding the concept of "learning outcomes," the affect and outcomes of learning by doing as presented in "The Teaching Hospital Model of Journalism", and the overarching concerns of diversity essential to learning within student media organizations as well as addressing their increasingly diverse audiences.

"Campus Media Advisors: Thoughts for the Future and Resources" by Mortimer W. Gamble and Patrick J. Sutherland details the media advisor's role as it adapts to a shifting media landscape—from more structured organizations that condition how learning occurs, to more independent and entrepreneurial modes—suggesting that media advisement is more important than ever. This essay ends with a short catalog of resources available to professional advisors. Those who have done this work will know that these and similar organizations are essential in keeping current about issues facing student media organizations, and in providing advice and networking for faculty advisors and professional managers who often have nobody to consult with on their campus.

Gregory Adamo's "Learning Outcomes and Student Media" examines the spillover of outcomes assessment from the formal curriculum to organizations traditionally considered "extracurricular." This essay addresses core issues about how learning in undergraduate media organizations is oriented, evaluated, and understood, and how such organizations subsequently, as venues for learning, are justified as part of the educational mission of institutions of higher education.

In "The Teaching Hospital Model of Journalism" Jacqueline Soteropoulos Incollingo examines the dominant model currently shaping the structure, implementation, and understanding of many undergraduate media organizations in the USA. Among other aspects, the essay examines how journalism educators and administrators define "The Teaching Hospital Model," how they apply it to their own teaching practices, whether the model replicates and/or supports traditional, legacy models of journalism, and the challenges of teaching within aspects of the model.

"Diversity, Learning and Media Organizations" by Allan DiBiase and Charles McKinney employs historical, personal narrative and case study perspectives in examining critical issues of diversity as they affect learning in undergraduate media organizations. The topics treated in this essay are of long-standing concern, but evidence shows relatively little progress in meaningfully addressing them, especially as new media outlets expand the compass of media consumers. How are students prepared through undergraduate learning experiences to meet these demands?

ACKNOWLEDGMENTS

The editors wish to thank the contributors to this volume for their collegial patience and perseverance throughout the process of bringing it to print. We also wish to acknowledge the work of others whose work could not be included: Judith Roberts, Austin Vining, Chris Delboni, Allan Richards, Edward Arke, the late Dan Reimold, and Lora Taub-Pervizpour. Special thanks to Linda Steiner for reading and advising about sections of the manuscript. The editors believe, as a whole, the book represents a unique and understudied aspect of college media organizations and we hope that the research and views expressed in this book lead to continued research and appreciation of the role of college media in students' education.

REFERENCES

Dunbar-Hester, C. (2014). *Low power to the people: Pirates, protest, and politics in FM radio activism.* Cambridge, MA: The MIT Press.

Hegel, G. (1977). *Phenomenology of spirit* (A. Miller, Trans.). New York, NY: Oxford University Press.

Jackson, D. (2012, March 8). In wired generation, students like paper for campus news. *Washington Times.* Retrieved from http://www.washingtontimes.com

Pérez-Peña, R. (2012, August 16). Student paper editors quit at University of Georgia. *New York Times.* Retrieved from http://www.nytimes.com

Scifo, S. (2015). Editorial. *Interactions: Studies in Communication & Culture, 6*(1): 3–7. Retrieved from http://www.intellectbooks.co.uk/journals

Student Press Law Center. (2014, November 3). Muting the airwaves: As colleges sell off their radio stations, student deejays grapple with their identities in the digital age. Retrieved from http://www.splc.org

Troop, D. (2011, June 19). What's eating college radio? *Chronicle of Higher Education.* Retrieved from http://chronicle.com/article/Whats-Eating-College-Radio-/127945

SECTION ONE

Theory

CHAPTER ONE

Experience AND Learning

ALLAN DIBIASE

> Because the need of preparation for a continually developing life is great, it is imperative that every energy should be bent to making the present experience as rich and significant as possible.
>
> —Dewey (1985a, p. 61)

INTRODUCTION

It's a commonplace that humans can learn through their experience. But less commonplace to examine exactly what is meant by the concepts of "experience" and "learning" and how the two are related in real-life situations. This chapter will examine these issues using the work of the American philosopher John Dewey (1859–1952). Dewey was born in New England prior to the Civil War and died in the midst of the Korean conflict. He wrote about and taught education, philosophy, and public issues extensively at many levels globally. He engaged a wide range of publics in debate on the issues of his time. Dewey was especially interested in change and how people respond to it. Over his long lifetime Dewey experienced upheavals in American social life associated with the expansion of railroads into the west, the advent of electricity, telephones, cars, antibiotics, radio, refrigeration, the atomic bomb, television, airplanes, women's suffrage, the great depression, many wars, and a host of other things too numerous to list. How did people respond to the rapid pace of change and what was its effect on their lives? How did they learn from these disruptive aspects we call change?

It's customary in our time to emphasize the changes wrought by the digital age. As for Dewey, such changes are unequally distributed and by their presence or absence create qualitative differences in how people experience the world in which they live and learn. Put differently, Dewey's lifelong engagement with the dynamics of change, his efforts to describe the nature of it and how it affected what we call "learning" is still relevant and applicable—perhaps due to the vast scope of change he lived through—even *more* relevant than what many of us have experienced in our lifetimes.

About one hundred years ago, in 1916, Dewey (1985a) published his most widely read and influential book, *Democracy and Education*. In it he methodically laid out his observations and thinking on the nature of social life and how people might productively learn through it. In this way, he unpacked the key concepts of "experience" and "learning" as he illustrated them by both example and thoughtful consideration. For Dewey it was critical to understand the meanings of experience and learning through the evolving situational conditions under which they occurred. It is no small thing that two of the media this book is most closely associated with—radio and television—were part of the profound changes Dewey experienced during his lifetime.

In *Democracy and Education* Dewey's first chapter is called "Education as a Necessity of Life." Within a few short paragraphs he focuses his discussion on the nature of "experience" in its broadest cultural or anthropological sense. Experience is a concept that encompasses the lived particulars, large and small of "life." For Dewey (1985a) life is "The continuity of any experience, through renewing of the social group …" (p. 5). Although we can discuss the lives of individuals (and have done so extensively in the historical record) the fact that life is maintained in association with other human beings is, for Dewey, a factual necessity and one not to be forgotten or ignored casually. How individuals live and function within a group, whether directly or through indirect relationship, whether positively or negatively, is for Dewey the essence of understanding their experiences. Perhaps even more important is how individuals perceive and utilize those understandings to inform their subsequent experiences.

Dewey (1985a) describes this as the notion of "continuity" in how human culture is transmitted (p. 5). In his view, continuity is maintained, most vitally, through communication: "The communication which insures participation in a common understanding is one which secures similar emotional and intellectual dispositions—like ways of responding to expectations and requirements" (1985a, p. 7). Shortly after this he wrote: "Not only is social life identical with communication, but all communications (and hence all genuine social life) is educative. To be the recipient of a communication is to have an enlarged and changed experience" (1985a, p. 8). Then, summarizing, he writes: "In final account, then, not only does social life demand teaching and learning for its own permanence, but the very process of living together educates" (1985a, p. 9). Later in his career, in a book called *Experience and Nature*, Dewey wrote: "Of all affairs, communication is the most wonderful" (Dewey, 1981, p. 132).

EXPERIENCE AND LEARNING

Using the above as a point of departure I want to look now at the concepts of "experience" and "learning" as they relate to the essays in this book. These essays are situated for the most part in undergraduate media organizations organized and provided to enhance the participants' experiences in learning about communication. Despite the many mediums of communication this book discusses, and the variety of organizational structures designed to deliver these experiences, a few things are common to all:

- All have a baseline concern with experiences involving the social transmission we call communication.
- All who participate in these organizations undergo specific qualities of experience associated with the life and function of these organization.
- It is a common aim of providing these organizations that the quality of experiences in these organizations can be identified by the participants, as well as others, as "learning."

What is it that we call "learning" that gets its definition through the experiences had in these organizations?

In an essay written in 1927–28 called "Progressive Education and the Science of Education" Dewey (1984) undertook to define "learning." Although the references in this article to "learning" are too numerous to cite here, one should be particularly interesting to readers of this book. In it Dewey zeroes in on how to identify learning: "A series of constantly multiplying careful reports on conditions which experience has shown in actual cases to be favorable and unfavorable to learning would revolutionize the whole subject of method" (p. 267). We can interpret "method" in this essay to mean the organization and life of a student media organization. In this sense, "learning" is best understood by careful observation of the participants' experiences in particular situations. For a more comprehensive understanding of Dewey's notion of "learning", see his 1937 essay called "What is Learning" (Dewey, 1987, pp. 238–241).

This view—about learning what learning *is* by careful observation—is echoed in a 1989 essay by James W. Carey called "Reconceiving 'Mass' and 'Media'" where he wrote:

> The objective of doing all this—of looking at the practices that organize communications, the concepts such practices presuppose, and the social relations they bring into existence—is a hermeneutic one: to try to find out what other people are up to, or at least what they think they are up to; to render transparent the concepts and purposes that guide their actions and render the world coherent to them; to extend the human conversation, to incorporate into our world other actors tending other dramas by comprehending what they are saying. (p. 85)

This way of characterizing the concept of "learning" can be distinguished from situations of "training" where very specific knowledge and procedures are to be taught, the outcomes of which are then to be enacted by the participants in some observable or testable fashion. Although this kind of training is part of any undergraduate media organization, it's usually the earliest phase of learning in a student's involvement. Moreover, in contrast to the training phase common to most media organizations, the larger goal—or in Dewey's parlance, the "end-in-view" (1981, p. 86)—is "learning to learn" a phrase coined by the educational philosopher Maxine Greene (1978, p. 3). It is our expectation that students will become self-activated and independent learners able to apply the dynamics involved in one learning situation intelligently to another. And, hopefully these modes will become part of their career and life in association with others no matter what that career or set of associations may be.

This process of observing and coming to both know and name what and how students are learning is hard compared to accepting a given formula for producing a predictable result and then perhaps testing for that result. The difficulties arise due to the necessity to capture an inclusive qualitative description of not just what and how much is learned, but the processes and dynamics involved in how it came to be. This provides us a focus on what Dewey (1985a) in *Democracy and Education* described in this way:

> The knowledge which comes first to persons, and that remains most deeply engrained, is knowledge of *how to do;* how to walk, talk, read, write, skate, ride bicycle, manage a machine, calculate, drive a horse, sell goods, manage people, and so on indefinitely. … Recognition of the natural course of development … always sets out with situations which involve learning by doing. (p. 192)

We could add to Dewey's list, learning how to write a lede, how to conduct an interview, learning to use the metrics of the newer forms of social media intelligently, learning how to edit your own and others' work. Dewey often used the example of learning to ride a bicycle since was a rather rich and full version of what an experience can be. It is this kind of indirect or "learning built-into an experience" or "natural course of development" that Dewey honored most and for which he urged teachers, facilitators, leaders, managers, directors, and administrators to give close observation and attention. It is fair to say that Dewey believed any kind of education and learning depended, in the long run, on the continuity within a given situation of successful (as well as unsuccessful) learning experiences. For him this meant the fullest understanding of the qualities of experiences that participants had.

So, we might ask, what is similar in the qualities of learning experiences associated with riding a bicycle and participating in a college media organization? Dewey frequently short-handed this as the fusion of intellectual and emotional aspects available in an experience. It's not hard to observe, identify, and describe what we might call intellectual and emotional insights and excitements that characterize the learning in both cases. However, in random places in his voluminous writing, Dewey alludes to many other aspects of experience all of which have the capacity to make an experience memorable enough to be utilized and called "learning." These aspects (views or senses of one's experience) observed in actual cases can serve as a qualitative description of that experience. Dewey believed that the fuller the experience, the more aspects evoked and appreciated, the greater the potential for learning. In a chapter of Dewey's 1932 book, *Ethics* (1985b) called "The Nature of the Criterion of Social Conditions," Dewey wrote directly to both the conditions and requirements of the learning associated with human development:

> The tenor of this discussion is that the conception of common good, of general well-being is a criterion which demands *the full development of individuals* in their distinctive individuality, not a sacrifice of them to some alleged vague larger good under the plea that it is "social." Only when individuals have initiative, independence of judgment, flexibility, *fullness of experience* can they act so as to enrich the lives of others and only in this way can a truly common welfare be built up. The other side of this statement, and of the moral criterion, is that individuals are free to develop, to contribute and to share, only as social conditions break down walls of privilege and of monopolistic possession. (p. 348, italics mine)

ASPECTS OF EXPERIENCE

In various essays and books Dewey evokes the following aspects of experience which we can call intellectual, physical, social, emotional, aesthetic, and spiritual. In this sense, the natural course of development of an experience is holistic in nature. By this I mean that the potential for learning is facilitated or impeded by all these aspects of a person's experience that come, or perhaps don't come into play in a given situation. This doesn't mean that these aspect-concepts for describing "a whole experience" are shared by everyone, nor are they a necessity. Rather, they are a useful (albeit, a reductive) way for us to grasp, observe, and describe something which is often treated simply as "experience" or as Dewey does sometimes more broadly, as "life."

Viewed in this way, the division of experience into these six aspects is arbitrary and not a claim about all human experience. It's simply a division by which it may be possible to better grasp the dimensions that characterize experiences and through which people learn. Dewey never "codified" these things in the way I'm doing here. Rather, he kept these aspects (except in short-hand reference in many places) integrated into the larger fabric of his presentations of ideas and thoughts concerning education and learning. As an educational philosopher Dewey was very careful about not erecting what appeared to be formal stages, formulas or grades for human development. However, in a 1941 essay called "The Philosophy of Whitehead" Dewey approvingly cites Whitehead on "experience" thus providing a concise and unified indication of what he meant when he used the phrase the "fullness of experience" (Dewey, 1985b, p. 348).

> We must appeal to evidence relating to every variety of occasion. Nothing can be omitted, experience drunk and experience sober, experience sleeping and experience waking, experience drowsy and experience wide-awake, experience self-conscious and experience self-forgetful, experience intellectual and physical experience, experience religious and experience sceptical, experience anxious and experience care-free, experience anticipatory and experience retrospective, experience happy and experience grieving, experience dominated by emotion and experience under self-restraint, experience in the light and experience in the dark, experience normal and experience abnormal. (Dewey, 1988, p. 126)

SIX ASPECTS

First considered in this essay is the intellectual aspect. I do not believe this is the aspect that comes first and naturally to most people. But usually (and increasingly) it becomes the foremost learning goal in situations considered "educational," and whether regarded explicitly or implicitly, it's then codified and enforced pervasively in many cultures. Among the benefits attributed to the intellectual are those of science and the resulting technologies. But clearly its benefits extend to all fields of human endeavor. In this essay, what's questioned is whether this aspect of experience need be paradigmatic of all learning? Regardless, it's not hard to imagine students whose involvement in media organizations is driven by an intellectual focus and we may even like and need that in our organizations. Some of these students may be those who have eventually written chapters for this book.

Second is the physical, kinetic aspect of experiences. This of course features prominently in learning to ride a bicycle and many other kinds of doing as was mentioned above. It's an aspect necessary for human life and one of the earliest ones about which humans seem to become aware. The eyes learn to focus. A cry gets attention. One learns to crawl and then walk. So this aspect of human experience is deeply engrained. Yet, often it has to be rediscovered after a period of devaluation as people grow up. Then it is often packaged not as a natural response to situations of life but as a kind of orderly and rational process, rationally derived; one that produces a desired result such as in "physical exercise." But what's important in this discussion is that the human physical body—its movements and actions in the long run, small and large—have long standing importance in how a person experiences the world and others. It's a necessary conditioning factor in the quality of all experiences including those of sitting and reading a book, raising an eyebrow, working on a computer, playing a sport, scratching one's head, or the sensation that what you want to say is "on the tip of your tongue."

The mainstream western tradition in education and its subsequent influence on what's considered learning has been to dichotomize the intellectual and the physical aspects of experience. This often results in them seeming oppositional rather than complementary and part of a whole response to any situation in which learning is a possibility. How we as learners come to see ourselves in this regard is of critical importance. To what aspects of our experience do we attribute our learning? In answering this question it's useful to see shades of distinction between the intellectual and the physical as integral to life and living.

The third aspect I want to consider is that of the social. As referenced earlier, Dewey considered this aspect of experience as essential to human existence. Moreover, Dewey (1984) considered the physical dimension of human life to be an equally essential factor (pp. 47–48). The two are intertwined necessities not just of human well-being and flourishing but to the basic survival and continuance of human life.

Another issue, perhaps more readily apparent in the western tradition, is that the social dimension or aspect of experience is frequently dichotomized with that of the individual. This almost always presents a distorted view of how humans live and learn. Of course, the philosophical debates about this are much more complex than we can get into here. It is possible to appeal in either direction—especially when studying and honing the individual or social viewpoint—but as in the discussion above about the intellectual and the physical, I believe the best view of any human situation involves both. It's fairly easy to see by looking around at how media organizations actually run, or university faculty meetings for that matter, and see how the tension between these two polarities plays out in the doing. People most often see themselves importantly as one or the other, but customarily less so as both. This false dichotomy distorts the reality of situational "givens" within which we all live and learn. To be human is to be social as fish are to water. This in no way precludes the realization and knowledge of a human self that has individuality.

Beyond Dewey's and my claims about the factuality of our social environment, it's hard to dispute that some students come to media organizations, or any form of student organization, through the social dimension and sense of their experience. Identity and belonging are often a factor. This need not be a primary or direct awareness but the barest form of common-sense tells us that we will inevitably work with students who have these needs and motivations as part of their interface with the world of others. Likewise, many will come with a strong interpretive sense that can only be called "individualism" (a concept very different from "individuality"). The particular dichotomy or tension between the social and individual is surely one that everyone involved with college media organizations has confronted.

Next, is the emotional aspect of experience which has often been discounted in favor of the more rational and intellectual that seems to be easier to chart, describe, and recommend as in the western tradition of philosophy. However, since the rise of psychology as a field of study over the last two centuries, separated from the traditions of western philosophy, emotions have been increasingly recognized as the coloring agents for all aspects of human experience (see Daniel Golman's (1995) *Emotional Intelligence*). This, in some cases has resulted in a field of study set off against the intellectual and rational. But again, as in the aspects of experience treated previously, a casual look around shows that dichotomizing emotions and the intellect is a false move. Although emotions can distort our experiences, they also are the point of departure for many human endeavors. And of course, in any process of learning where there is a developmental curve, emotional excitement and discovery often lead the way. One can only think about how passionate students can be about music at a college radio station. In various chapters in this book you can read about and sense this powerful aspect of experience and how it contributed to a memorable and life-long learning experience.

The fifth aspect of experience we can look at is the aesthetic and creative dimension. The ability "to make" is a powerful capacity of human beings. So in most people's experiences there is an identifiable aspect associated with making or creating with others or though more individually sensed processes of thought and action. This is a well-known fact within both the arts and the sciences. In an organizational structure, college media or otherwise, just as some involved need access to and a focus on the intellectual, others may need greater availability for creative endeavors and doing, with the caveat as above that the two are not mutually exclusive. We see this when students devise new programs and formats, create new web-designs and templates for media delivery; and when they write and express themselves in addressing topics, themes and issues in print and other media presentations.

The sixth aspect of experience I want to consider is the spiritual. I believe for most people there will be some aspects of their experiences that go beyond explanation or description. Although this seems not to be true for everyone, it's still seems useful to have a category that accommodates unarticulated and hard to observe aspects of our experience. By putting it this way I mean to disassociate a direct link between spirituality and formal religion, although, it may be religious belief that is the spiritual aspect of experience for some people. In a media organization, a person may find identity through social belonging and a kind of pleasure or excitement that comes with that. This may not be distinctly articulated by that person and that's what I consider then to be a spiritual aspect of a person's experience. Might this not explain how some students become so passionate excited and committed to their involvement in a media organization?

These six aspects by which a person can articulate a view of their experience, as mentioned above, are an arbitrary means of better grasping what we mean when we appeal to experience or doing in learning. Any programmatic effort that appeals to "learning by doing" can be judged and evaluated by its openness to these aspects. They are features that can contribute to what Dewey called "the natural course of development." It's not that all these aspects must be present or that they actually are. But the possibility of people relating to their experience through these aspects, I believe, provides the greatest opportunity for learning to occur not through a single aspect, but, in the diverse and plural ways this model suggests. In a plural and diverse society, recognition of this becomes all the more important. Once this is allowed and honored, then the facilitator's task becomes to monitor the available experiences being had so the organization can remain "tuned up" to the best learning dynamics that can be achieved.

EXPERIENCE IN OUR TIME AND PLACE

The genius of "extra-curricular" organizations at any level of formal education is they are, in part, ways of highlighting and utilizing the most formative modes of learning that many people get in an entire lifetime. This learning is the learning associated with family life. Family life is largely a sustained period of unarticulated holistic learning. Although, according to some accounts, this arena for learning is much attenuated in many modern cultures, it is, in its ideal form, and practice as well, still the gold standard for "a natural course of development." When Dewey wrote "the very process of living together educates" (1985a, p. 9) I think most of the writers of essays in this book have had this experience within and through an undergraduate media organization even if not in an especially memorable family situation. Although it is not exactly "family life" or "living together" (despite the couches in many basement radio stations) it is still, in varying degrees and modes, a form of life lived together directed toward a common purpose.

The most disruptive force the digital revolution has had on traditional media organizations—which were, and still are, largely based on face-to-face, communication and interaction—is the attenuation of whole, face-to-face, shared experiences had in common physical space. The trend toward free-lance forms of journalism, encouraged by the enormous variety of platforms for packaging and delivering media (compared to the past) has produced increasingly different modes communication and transaction in the production of media. These virtual, non-physical situations often eliminate or denature aspects of the learning experience, in particular face-to-face social interaction, in favor of more independent modes of production. This is a foundational change worthy of the kind of close observation and study that Dewey recommended. What is being gained and what is being lost in this change or shift?

In Dewey's *Democracy and Education,* in Chapter 7, "The Democratic Conception in Education" (1985a) Dewey develops his ideas about democracy as a social arrangement for living and learning together. Part way through the chapter Dewey observes:

> Every expansive era in the history of mankind has coincided with the operation of factors which have tended to eliminate distance between peoples and classes previously hemmed off from one another. Even the alleged benefits of war, so far as more than alleged, spring from the fact that conflict of peoples at least enforces intercourse between them and thus accidentally enables them to learn from one another, and thereby to expand their horizons. Travel, economic and commercial tendencies, have at present gone far to break down external barriers; to bring peoples and classes into closer and more perceptible connection with one another. It remains for the most part to secure the intellectual and emotional significance of this physical annihilation of space. (p. 92)

Today, in the world that's developed in a certain way, it's possible, even pervasive in some instances, to communicate in a virtual versus a physical space. This is a profound change in the basic situational conditions, the spaces in which communication happens. But it's not that we are further apart, it's that we are so immediately accessible to each other, and often chiefly through the filters of various media. I think Dewey in the above passage meant to give a balanced appraisal of this kind of change versus mindlessly celebrating it, for it comes with many benefits and many risks as we are learning now: From real life invasions of privacy, to terrorist recruitment, to the exploitation of children, and a new kind of anxiety caused by a constant bombardment of breaking news. Have we asked what kind of communications facilitate human flourishing versus modes of interaction and communication which are merely symbolic of it?

What is *also* being lost, as colleges and universities sell off their radio stations and transform their media organizations into adjuncts to the classroom, are enduringly deep and unique spaces for learning. This happens as the drift of higher education slowly moves toward a more uniform mode of classroom organization and teaching. One of the more current manifestations of this is the growing requirement that even informal organizations must state learning outcomes. And what good are projected "learning outcomes" without learning assessments and tests? It's not that extracurricular and media organizations don't have a purpose, goals or even a mission within a university and that these cannot be evaluated. But whether a more holistic model of learning that characterizes media organizations can be lumped into the current trend of testing in classroom teaching is, in my view, a disturbing move, if only for the fact that it erodes diversity of approaches to learning and somewhat blindly caters to classroom practice.

In my own experience of teaching graduate classes in education, through the kinds of close observations discussed in this chapter, I was able to evaluate and adjust the delivery of courses ranging from those totally online and taught around the globe; to the more traditional locally based classroom teaching weekly, face-to-face meetings; to a variety of hybrid teaching modes conducted partly online, partly in a classroom and sometimes in outdoor situations or in foreign counties. Adjusting the mix of how people worked independently as readers and interpreters and then shared their experiences online and in face-to-face meetings was a constant challenge. All in all, the enacted modes of "living together" for a particular purpose—even if we were "living together" in an online template—were under constant observation and evaluation for the quality of communication and resulting learning. This itself was the model being enacted or recommended *to* the educators who took these courses. I think there's some possible cross-over from this kind of contemporary experience on the formal side of education to those organizations which always have represented alternative, extra-curricular forms of learning that are freely elected by participants. It's a most interesting time to be alive and facilitating learning.

This observation obviously involves some sort of trade-off between the positives and negatives in the immediacy of shared, holistic face-to-face experiences, and the positives and negatives of immediate but potentially more reflective experiences offered by the digital age we live within. Put differently, working online with others *can* involve aspects of the types of experiences discussed above. Whatever the case, careful observation of actual learning experiences and how they affect the quality of learning is essential.

CONCLUSION

There's no doubt that the learning-education that refuses to be measured and predetermined is more difficult to understand and describe than many other types. The best we can do is give demonstrations and carefully examine their occurrence acknowledging that every situation of natural learning involves a specific location and individuals that make the dynamics involved unique. A prominent characteristic of such learning is that it's the *least* of "participants being told what to do and following" and the *most* of them learning to think and act independently and critically. This is the kind of learning that comes out of situations where a respectful freedom and creativity is understood and cultivated. These are situations that become the seedbeds for more vital forms of journalism. Situations that make contributions to larger common, and even new, ways of life and living that result from the enactment of these same dynamics in the production of the media.

The sad fact that most universities and colleges in the USA today are no longer themselves vital enactments of these notions requires that media organizations of any type continue to offer a critical alternative to the status quo. How to do this through organizations allied to higher education is a core challenge to the field considering, for example, situational elements of digital disruption that all too easily prompt students to join in parroting the metrics of user likes. More broadly, the fact that the facilitation of learning *how to do* communications in any mode or form involves processes that must be endlessly reinvented as social and cultural conditions evolve. This chapter has proposed a framework useful for appreciating and observing such dynamics in a finer grained way. One that offers prospects for justifying and supporting the unique opportunities that undergraduate media organizations afford for learning.

REFERENCES

Carey, J. W. (1989). *Communication as culture: Essays on media and society.* New York, NY: Routledge.

Dewey, J. (1981). In J. A. Boydston (Ed.), *The later works, 1925–1953, vol. 1, Experience and nature.* Carbondale, IL: Southern Illinois University Press.

Dewey, J. (1984). In J. A. Boydston (Ed.), *The later works, 1925–1953, vol. 3, Essays.* Carbondale, IL: Southern Illinois University Press.

Dewey, J. (1985a). In J. A. Boydston (Ed.), *The middle works, 1899–1924, vol. 9, Democracy and education.* Carbondale, IL: Southern Illinois University Press.

Dewey, J. (1985b). In J. A. Boydston (Ed.), *The later works, 1925–1953, vol. 7, Ethics.* Carbondale, IL: Southern Illinois University Press.

Dewey, J. (1987). In J. A. Boydston (Ed.), *The later works, 1935–1937, vol. 11, Essays.* Carbondale, IL: Southern Illinois University Press.

Dewey, J. (1988). In J. A. Boydston (Ed.), *The later works, 1925–1953, vol. 14, Essays.* Carbondale, IL: Southern Illinois University Press.

Greene, M. (1978). *Landscapes of learning.* New York, NY: Teachers College Press.

CHAPTER TWO

College Media AS AN "Extracurricular Activity"

Historical and Theoretical Perspectives

MELONY SHEMBERGER

INTRODUCTION

Extracurricular activities help the student become the creator rather than the consumer of knowledge (Katz & Hartnett, 1976). Studies have documented well the benefits of extracurricular activities in education. Although initially regarded as a distraction, extracurricular activities are deemed an essential part of the college experience (Conway, 2009). The same is true for college campus media.

Working in the structured environment of college media offers many benefits. Students learn the role that public feedback plays in news and content, experience deadline pressures, and learn from mistakes—all of which help them to prepare for a job, either in the news profession or elsewhere. In addition, being involved in college media can help students find their identities, socially and professionally.

All of these outcomes can be described through a number of historical and theoretical perspectives, primarily those centered in student development and involvement frameworks. Student development theories can assist professionals, educators, and administrators to understand better the student populations with whom they work. Astin's theory of student involvement (1984), for example, helps to explain the practice of extracurricular activities in American higher education.

This essay will explore college media as an extracurricular activity through a number of theoretical lenses, including that of Astin's student involvement. This chapter is designed to help campus media advisers, as well as any educator who works with student communicators, apply a theoretical foundation to create effective campus media learning environments. In addition, a historical perspective of extracurricular involvement in the American higher education system is provided, along with the historical development of college media participation.

HISTORICAL BACKGROUND

Before examining college campus media through a theoretical lens, it is interesting to discuss the topic historically. As for student college newspapers, it is difficult to tap one college newspaper as America's oldest paper or declare the first one in the United States. That's because changes in the student paper's name or publication frequency make the special distinction tough to make. For example, a student newspaper began publishing at Dartmouth College in 1799. It started as the *Dartmouth Gazette*, but later became *The Dartmouth* sometime in the 1800s (Reimold, 2013).

Regardless of which college campus paper was the first, longest or oldest, these publications have been a part of the collegiate tradition since the American higher education system developed. Over time, the idea of the four-year residential experience impacted the belief in what college or university life was about. Much research has focused on the collegiate experience in which the paradigm was a residential setting complete with extracurricular activities (Cohen, 1998). This environment became synonymous with what college was supposed to be—that is, student life. By the 1930s, a student personnel point of view had been formulated into the belief that a college was responsible for all aspects of a young person's life, including emotional and psychological characteristics as well as learning and cognitive development (Cohen, 1998).

This belief also extended to extracurricular activities, including sports, drama clubs, debating societies, fraternities, and religious observances. The extracurriculum became a way for students to participate in comparable activities in the outside world. This was the era in which organized sports and organized debating clubs grew into intercollegiate activities (Cohen, 1998). Among academic extracurricular activities are journalism or media activities, such as participating in the production of a yearbook, working for a newspaper or TV/radio station, taking a journalism class, or being involved with some sort of online or web journalism instruction. Like other academic activities, journalistic activities have been found to have positive effects on student academic performance.

Research on the effects of these extracurricular activities involving communication venues supports the idea that they have positive effects on student academic performance (Becker, Han, Wilcox, & Vlad, 2014). Generally throughout high school and into college, students who participated in journalism activities have been shown to score higher in reading and comprehension (Becker et al., 2014).

The next section of this chapter focuses on the outcomes of extracurricular activity participation. Other research has demonstrated that high school seniors and college freshmen who participated in school newspaper and other journalism activities had greater grade-point averages and committed fewer errors (Dvorak, 1990, 1998; Dvorak & Choi, 2009; Dvorak, Lain, & Dickson, 1994). Scholarly inquiry also has shown that the best predictors of success in the job market are participation in college media, internships and specific career selection (Becker, Kosicki, Engleman, & Viswanath, 1993; Becker, Lauf, & Lowrey, 1999).

OUTCOMES GENERATED BY EXTRACURRICULAR ACTIVITIES

Students who work in campus media have opportunities to gain a number of benefits. The most obvious one is skill development that will prove useful for the job market. However, there are other positive outcomes that result by participating in campus media or other extracurricular activities. Participation in student activities has a positive effect on grades, no matter the various background variables that exist (Gamp, 1990). Other researchers such as Huang and Carlton (2003) have found the same relationship in their studies. One explanation of these findings is that students are more likely to do well when they engage in extracurricular activities because they become committed to the school's goals and identify with the school's culture (Keenan, 2010).

In addition to academic success, many studies have shown that an important outcome of extracurricular activities is skill acquisition (Eccles & Barber, 1999; Gilman, 2004; Mahoney & Stattin, 2000; Zaff, Moore, Romano, & Williams, 2003). Job seekers and recruiters believe that being a part of an extracurricular activity is indicative of the quality of their skill set (Conway, 2009). In fact, "participation in student government, the school paper, and debating teams were most likely to relate to meaningful performance criteria" (Howard, 1986, p. 545).

Participating in extracurricular activities also results in career success through a number of psychological outcomes. Successful careers need "well-developed cognitive abilities, interpersonal competence and motivation" (Chickering, 1994, p. 51). Extracurricular activities also have been tied to outcomes in improved mental health (Feldman & Matjasko, 2005; Gilman, 2004).

Extracurricular activities can lead to outcomes involving development of social ties. Social connections obtained through involvement in student media can be as important or more important than any other type of relationship, especially when it comes to professional networking and employment. Granovetter (1974) concluded that personal contacts were considered superior to formal methods of job location and that even weak ties were the most successful means of finding a job.

THEORETICAL FRAMEWORKS

The outcomes and benefits of participation in extracurricular activities can be linked to a number of theoretical frameworks. Because college students are complex, multifaceted individuals, it is best not to declare any one theory as an exclusive fit. It is important for campus media advisers, who are leadership educators in a student newsroom, to integrate a number of theories, models, perspectives, and tools.

What follows are brief summaries of selected theories often applied in educational research, including student participation in extracurricular activities. These are not comprehensive descriptions, but they offer sufficient background for the college media adviser to consider and apply to his or her own campus media environment.

STUDENT INVOLVEMENT

An examination of college media as an extracurricular activity can be rooted significantly in Astin's student involvement theory. Educational research supports this theoretical framework for student involvement. Astin began constructing a student development theory based on student involvement in 1984, although the involvement theory has its origin in a longitudinal study of college dropouts that sought to identify factors in the college environment that significantly impact the student's persistence in college (Astin, 1975). He contends that student involvement is a condition for student retention. Based on his involvement theory, Astin (1984) believes that the more students are socially involved with campus life, the more likely they will persist and graduate.

Although this chapter will review other theories that are applicable to the discussion of college media as an extracurricular activity, the theory of student involvement is perhaps the most appropriate for several reasons. First, the theory is simple to understand, use, and explain. Second, the theory is comprehensive; it explains most of the empirical knowledge about environmental influences on student development that researchers have learned over time. Third, Astin's theory embraces principles from other theories. Fourth, student involvement theory can be used by researchers, college administrators, and faculty to help them build more effective learning environments. Researchers and practitioners have used it to develop programs, modify curriculums, make administrative decisions, and conduct research.

Involvement can be defined as an investment of physical and psychological energy that occurs along a continuum and has both quantitative (e.g., time spent) and qualitative (e.g., focus or depth) features (Astin, 1996). An "involved student" is one who "devotes considerable energy to studying, spends much time on campus, participates actively in student organizations, and interacts frequently with faculty members and other students" (Astin, 1996, p. 518). Therefore, student involvement has a strong behavioral component. It is not so much what the individual thinks or feels, but what the person does, how he or she behaves, that identifies involvement (Astin, 1984).

One of the characteristics of the student involvement theory is its foundation in certain principles. These principles can be applied to understanding extracurricular activities and even building a particular program. Astin's theory (1984) has five principles:

1. There is investment of physical and psychological energy in various objects;
2. Involvement occurs along a continuum in which students demonstrate different degrees of involvement in the object. A single student will show different degrees of involvement in different objects at different times;
3. Involvement has both quantitative and qualitative features;
4. The amount of student learning and personal development associated with any educational program is directly proportional to the quality and quantity of student involvement in that program; and
5. The effectiveness of any educational practice is directly related to the capacity of that practice to increase student involvement.

These five tenets of the student involvement theory can be applied to campus media learning environments. Beginning with the first principle, experience in campus media would be the defined object. Students in the same campus media environment will experience varying levels of involvement, according to the second principle. For the third principle, the extent of a student's involvement in campus media can be measured quantitatively (how many hours the student spends working on a story or in the newsroom) and qualitatively (whether the student reflects on his or her experiences and applies lessons learned in the classroom). What students learn from a campus media program depends on the ability of the program to involve students, based on the fourth and fifth principles.

PEDAGOGICAL THEORIES

The student involvement theory discussed previously relates to other traditional theories that will be discussed in this chapter, primarily those that focus on curriculum. Educators might find that the student involvement approach provides a link between various pedagogical theories and their outcomes. The next three pedagogical theories to be introduced cannot work independently. In other words, the student may or may not succeed in a particular set of courses simply because he or she was exposed to them. The involvement theory serves as a conceptual component for these three pedagogical theories (Astin, 1984).

Subject-Matter Theory

Also known as the content theory, the subject-matter theory of pedagogy has a strong emphasis on course content and is favored among college faculty. According to this theory, student learning and development depend on the right subject matter. Students learn by attending lectures, staying current on reading assignments and completing assigned work.

To some extent, this theory reflects the traditional way of student learning, but this approach can limit some students. Applying this theory favors highly motivated students; however, students who read slowly or have no interest in the subject matter of a course will not benefit from this approach.

In addition to course content, the content theory emphasizes specialized knowledge. As such, this framework encourages specialization of faculty interests by pairing faculty to courses based on their scholarly expertise and pedagogical ability.

Resource Theory

According to the resource theory of pedagogy, if adequate resources are used effectively in a setting, student learning and development will occur. Resources include laboratories, libraries, financial aid, counselors, faculty members, and a host of other facilities, programs, and personnel. The resource theory also includes the belief that high-achieving students are a resource. College administrators apply this belief to invest considerable financial resources to recruit high-achieving students. The problem with this approach is its attention on building resources without investigating whether students really will need or will use the resource.

Individualized Theory

The individualized theory pertains to the elective course content in the college curriculum. Most college curricula represent both the subject-matter and individualized theories, since students must take certain required courses but can take elective courses to help meet degree requirements. This theoretical approach is also associated with instructional techniques such as self-paced instruction. However, while this theory is appealing to educators, the downside to using this framework fully is that it is difficult and expensive to put it into practice because each student would require significant individualized attention.

The pedagogical theories discussed in this section add a rich and needed layer to the student involvement theory. If a college media adviser or instructor desires to build a campus media learning environment based on the student involvement theory, then the adviser should consider including these pedagogical theories or other frameworks to support the foundation of the involvement theory.

STUDENT DEVELOPMENT

Extracurricular activities help to build a positive self-image, reduce negative behavior, allow students to learn and continue to develop social skills, and provide mentoring possibilities between advisers and older leaders (Zaff et al., 2003). It also is believed that experiences outside the classroom are "extensions" of what is learned inside the classroom (Kennedy, 2008). The benefits of extracurricular activities extend far past formal education and have been shown to increase job quality and encourage greater participation in political processes (Eccles & Barber, 1999).

The theory of student development explains "the ways that a student grows, progresses, or increases his or her developmental capabilities as a result of enrollment in an institution of higher education" (Rodgers, 1990, p. 27). This perspective enables educators to "proactively identify and address student needs, design programs, develop policies, and create healthy environments that encourage positive growth in students" (Evans, Forney, & Guido-DiBrito, 1998, p. 5). This development occurs after a challenge, because when students are faced with new challenges, they respond in some manner (Sanford, 1966). If the challenge is resolved positively, this new response adds to the development of the student's identity. If the challenge is resolved negatively, then stress or anxiety may occur, and that could lead to a regression of the student's development. The amount of support available to the student when these challenges surface is crucial to whether the student's experience will be positive or negative.

Astin (1984) believed that his theoretical basis of student involvement contained a common construct in psychology—motivation. This is where the involvement theory can complement the student developmental theory just discussed. The theory of student involvement is different from the development theory, which focuses primarily on developmental outcomes or the "what" of student development. The involvement theory deals more with the behavioral mechanisms or processes that facilitate student development. In other words, the involvement theory addresses the "how" of student development.

An aspect of the student development theory with respect to campus media is professional or career development. Professional identity development is achieved during the process of professional socialization in which an individual adopts the values, skills, and attitudes needed to join a given society, group or organization (Merton, 1957). Socialization generally occurs in two major stages. In the first phase known as anticipatory socialization, the individual learns about the organization and decides to join after becoming assimilated to the organization's norms, values, and attitudes. The individual then enters the second phase, referred to as role continuance, and experiences the socialization processes that will influence his or her decision whether to remain in the organization and adopt the values, attitudes, and beliefs of the culture (Tierney & Rhoads, 1994).

Related to professional identity is social identity development, which describes "the ways in which individuals construct their various social identities, namely, race, ethnicity, gender, and sexual orientation, and the intersection of these multiple identities" including "social class, ability and disability, and religion" (McEwen, 2005, p. 13). Some scholars have included social identity in student development models because it shares in common many of the same developmental components found in other social development theories. These various explanations of social identity development allow for a deeper understanding of how an individual progresses from the least sophisticated view of himself or herself in the world, to one that is much more complex. This view goes beyond the individual's need for self-identification; it extends in the larger scope of other's needs and identities.

In the context of college media, students who join a campus publication or broadcast program are made members of an immediate community who value the freedom of information and the watchdog function of the press. These students are directly socialized into the role of a reporter and are given preparatory socialization into a journalist's life and future career. Furthermore, students who continue working in campus media will construct their identities and build them further when they reach the profession after graduation in the industry.

LEARNING THEORIES

The college media environment fosters learning. Learning theories offer much to say about how people learn and about how they learn how to learn. One of the widely known learning theories in the collegiate environment is that of David Kolb, who defined learning as "the process whereby knowledge is created through the transformation of experience" (1984, p. 38). According to Kolb's work, people process information through either of two methods: reflective observation or experimentation. Kolb describes learning as a four-step cycle that includes these elements: concrete experience, reflective observation, abstract conceptualization and active experimentation. All learning processes include some kind of concrete experience that leads to a period of reflective observation. From this reflection, students begin to conceptualize their learning in abstract terms and enter into a period of active experimentation. During this step, they assess the generalizations that resulted from reflection. This experimentation then brings forth another cycle in which more concrete experiences occur.

Theories such as Baxter Magolda's epistemological reflection model (1992) and Magolda and King's learning partnership model (2004) identify environments and conditions that promote students' self-authorship, or how one's construction of knowledge evolves from external to internal definitions of self.

Kegan and Lahey's plateaus in mental complexity describe the process of how individuals make meaning of their experiences (Kegan & Lahey, 2009). Meaning-making originates in the socialized mind that focuses on the expectations of others, shifts to the self-authoring mind that offers its own ideas, and finally becomes the self-transforming mind that seeks ideas and solutions.

Although the student involvement theory differs from the student development theory, the two make a highly useful and powerful combination when constructing or overhauling a campus media program. In fact, college media advisers must be alert to these theories as social media and digital technology are vital in today's communication and media landscape. For an effective program, advisers should conceptualize what students are expected to learn, or the outcomes, by participating in the campus media environment. This would allow advisers to engage the student development theory in this part of the planning. Connecting the student involvement theory, advisers then can evaluate why students should achieve the desired developmental outcomes.

CONCLUSION

College newspapers and other campus media programs should function as laboratories of innovation and courage. To achieve such an effective setting, various theories can prove helpful for college media advisers and instructors who want to build effective learning environments for students in campus newsrooms and production settings. There are more theories that could be included in this discussion, but this chapter encapsulated a few models most apropos in the context of college campus media as extracurricular activities. Marsh and Kleitman wrote that "whereas it is unlikely that one design or approach will ever provide a definitive solution … accumulating evidence using a variety of different approaches provides increasingly stronger support for the generalizability of the benefits of participation in extracurricular activity" (2002, p. 506). Theories can be thrown at situations and settings without proper thought to what their use would mean, but that would prove ineffective. Therefore, it is important that educators combine multiple theories, tools, models, and other principles they know or have readily available in their arsenal.

Working with a campus media outlet's leadership team—editors, directors, producers—will help advisers to create meaningful opportunities and encourage other students to participate. The theories discussed in this chapter can provide some suggestions for these experiences and outcomes. These guiding frameworks can help instructors and administrators assess their own activities in terms of their success in encouraging students to become more involved in the college experience.

REFERENCES

Astin, A. W. (1975). *Preventing students from dropping out.* San Francisco, CA: Jossey-Bass.

Astin, A. W. (1984). Student involvement: A developmental theory for higher education. *Journal of College Student Development, 40*(5), (September/October 1999), 518–529.

Astin, A. W. (1996). Involvement in learning revisited: Lessons we have learned. *Journal of College Student Development, 37*, 123–134.

Becker, L. B., Han, J. Y., Wilcox, D., & Vlad, T. (2014). The effects of pre-university study of journalism on entry to the job market. *Journalism & Mass Communication Quarterly, 91*(2), 344–356.

Becker, L. B., Kosicki, G. M., Engleman, T., & Viswanath, K. (1993). Finding work and getting paid: Predictors of success in the mass communications job market. *Journalism & Mass Communication Quarterly, 70*(4), 919–933.

Becker, L. B., Lauf, E., & Lowrey, W. (1999). Differential employment rates in the journalism and mass communication labor force based on gender, race, and ethnicity: Exploring the impact of affirmative action. *Journalism & Mass Communication Quarterly, 76*(4), 631–645.

Chickering, A. (1994). Empowering life-long self-development. *The Journal of the National Academic Advising Association, 14*(2), 50–53.

Cohen, A. (1998). *The shaping of the American higher education system: Emergence and growth of the contemporary system.* San Francisco, CA: Jossey-Bass.

Conway, A. (2009). An investigation into the benefits of extracurricular activities like clubs and societies to students and colleges: Are these benefits evident in the opinions and perceptions of staff and students in DIT? Retrieved from https://creativitythecityandtheuniversity.files.wordpress.com/2012/03/universitiessocietiesclubs20102.pdf

Dvorak, J. (1990). College students evaluate their scholastic journalism courses. *Journalism & Mass Communication Educator, 45*(1), 36–46.

Dvorak, J. (1998). Journalism student performance on Advanced Placement Exams. *Journalism & Mass Communication Educator, 52*(3), 4–12.

Dvorak, J., & Choi, C. (2009). High school journalism, academic performance correlate. *Newspaper Research Journal, 30*(3), 75–89.

Dvorak, J., Lain, L., & Dickson, T. (1994). *Journalism kids do better.* Bloomington, IN: ERIC Clearinghouse.

Eccles, J., & Barber, B. (1999). Student council, volunteering, basketball or marching band: What kind of extracurricular activity involvement matters? *Journal of Adolescent Research, 14*(1), 10–43.

Evans, N. J., Forney, D. S., & Guido-DiBrito, F. (1998). *Student development in college: Theory, research, and practice.* San Francisco, CA: Jossey-Bass.

Feldman, A., & Matjasko, J. (2005). The role of school-based extracurricular activities in adolescent development: A comprehensive review and future directions. *Review of Educational Research, 75*(2), 159–210.

Gamp, W. (1990). Participation in student activities and achievement: A covariance structural analysis. *The Journal of Educational Research, 83*(5), 272–278.

Gilman, R. (2004). Structured extracurricular activities among adolescents: Findings and implications for school psychologists. *Psychology in the Schools, 41*(1), 31–41.

Granovetter, M. (1974). *Getting a job: A study of contacts and careers.* Cambridge, MA: Harvard University Press.

Howard, A. (1986). College experiences and managerial performances. *Journal of Applied Psychology, 71*(3), 530–552.

Huang, C., & Carlton, B. (2003). The relationships among leisure participation, leisure satisfaction, and life satisfaction of college students in Taiwan. *Journal of Exercise Science and Fitness, 1*(2), 129–132.

Katz, J., & Hartnett, R. T. (1976). Recommendations for training better scholars. In J. Katz & R. T. Hartnett (Eds.), *Scholars in the making: The development of graduate and professional students* (pp. 261–280). Cambridge, MA: Ballinger.

Keenan, L. (2010). The effect of extracurricular activities on career outcomes: A literature review. *Student Psychology Journal, 1*, 149–162.

Kegan, R., & Lahey, L. (2009). *Immunity to change.* Boston, MA: Harvard Business Press.

Kennedy, R. (2008). The need for high school extracurricular activities. *Coach Athletic Director, 78*(4), 38–39.

Kolb, D. A. (1984). *Experiential learning: Experience as the source of learning and development.* Englewood Cliffs, NJ: Prentice Hall.

Magolda, B. (1992). *Knowing and reasoning in college: Gender-related patterns in students' intellectual development.* San Francisco, CA: Jossey-Bass.

Magolda, B., & King, P. M. (2004). *Learning partnerships: Theory and models of practice to educate for self-authorship.* Sterling, VA: Stylus.

Mahoney, J., & Stattin, H. (2000). Leisure activities and adolescent anti-social behavior: The role of structure and context. *Journal of Adolescence, 23*, 113–127.

Marsh, H., & Kleitman, S. (2002). Extracurricular school activities: The good, the bad, and the non-linear. *Harvard Educational Review, 72*(4), 464–514.

McEwen, M. K. (2005). The nature and uses of theory. In M. E. Wilson & L. Wolf-Wendel (Eds.), *ASHE reader on college student development theory* (pp. 5–24). Boston, MA: Pearson Custom Publishing.

Merton, R. K. (1957). *Social theory and social structure.* Glencoe, IL: The Free Press.

Reimold, D. (2013, September 18). America's oldest college newspaper? At least 8 papers claim the title (sort of) [Web log post]. Retrieved from http://www.collegemediamatters.com/2013/09/18/americas-oldest-college-newspaper-at-least-8-papers-claim-the-title-sort-of/#respond

Rodgers, R. F. (1990). Recent theories and research underlying student development. In D. G. Creamer & Associates (Eds.), *College student development: Theory and practice for the 1990s* (pp. 27–79). Alexandria, VA: American College Personnel Association.

Sanford, N. (1966). *Self and society.* New York, NY: Atherton Press.

Tierney, W. G., & Rhoads, R. A. (1994). *Enhancing promotion, tenure and beyond: Faculty socialization as a cultural process.* Washington, DC: George Washington University.

Zaff, J. F., Moore, K., Romano, A., & Williams, S. (2003). Implications of extracurricular activity during adolescence on positive outcomes. *Journal of Adolescent Research, 18*(6), 599–630.

CHAPTER THREE

College Media AND Leadership Training

Conceptualizing Student Leadership

LAURA K. DORSEY-ELSON

INTRODUCTION

Now more than ever there is a fervent interest in the study and practice of leadership (Clawson, 2011; Goethals, Burns, & Sorenson, 2004; Jackson & Parry, 2011; Norhia & Khurana, 2010; Northouse, 2015). Questions of what leadership is, what traits, behaviors, practices, and contexts result in the best forms of leadership and where the uncharted territories of leadership as a process remain, are among the key areas of research, theory building, and application in this exceptionally interdisciplinary field.

Even with over 50 years of leadership study in a variety of disciplines, there has been only a limited focus on leadership within the field of communication studies. The greatest amount of leadership and communication scholarship has occurred within the fields of interpersonal and organizational communication with much of the emphasis on how those who singularly succeed as leaders can utilize various communication strategies that support both individual and organizational success (Barrett, 2011; Fainhurst & Grant, 2010; Gamble & Gamble, 2013; Gardner, 2003; Miller 2015; Modaff, Butler, & Dewine, 2012; Vires, Bakker-Pierpr, & Aastenveld, 2010). While these are important contributions, there remains a need to examine more thoroughly, the nexus of leadership and communication and the unique forms and applications that come with it.

In an effort to examine better characterize, this area of study, this chapter examines the leadership stories that lie within the specific context of college media. The important role of the student-run and often faculty-advised media organizations should not be overlooked. Many students of mass communication with a specific interest in print, radio, television, Internet, and broadcast careers have found vital exposure, experience, and training in this context. Moreover, these student media organizations offer real-world professional experience that holds great value for career preparation. Oftentimes, the roles, rules, and responsibilities in college media organizations mirror those that exist in professional media environments, and with this comes a myriad of implicit leadership skill development for a student. Students also can be offered explicit leadership training via actual workshops, skill building sessions, etc. by the faculty advisors who oversee these organizations, thus furthering the leadership experience that the college media context offers. These types of leadership development opportunities are important to understand in the overall study of leadership and have the potential of moving communication to a more central focus in general.

This chapter provides a concise and up-to-date review of leadership theory and how it has been implemented via training programs in organizational contexts both similar to and apart from what the college media organizational context may provide. It will then offer the results of analyzed interview data from both college media managers and former college media students who have a valuable perspective on the impact of student leadership training within their specific university-based media organizations. This data will be thematically categorized within existing leadership development theory and potential new lines of thought will be suggested. The chapter concludes with some suggestions for further leadership inquiry within the field of mass communication and beyond.

THE EVOLUTION OF LEADERSHIP STUDIES

Leadership Theory

The study of leadership has been long-term and gradual, moving from a simple to a more complex phenomenon. Much of the most recognized leadership study work has been in the area of leadership theory. Goethals et al. (2004) provide a succinct, but comprehensive overview of this area of leadership studies. For the purposes of this chapter, three of the most relevant approaches to communication-based leadership research will be reviewed.

The first approach that has relevance to the field of communication is "Situation and Contingency Leadership Theory" (Hershey & Blanchard, 1969). This model of leadership conceptualizes the relationship between the situation and the leader and categorizes leaders into one of two groups: those who are more task oriented and those who are more relationship oriented. A scale (high, medium, low control) is used to determine which type of situation in which a leader will perform more effectively. Specifically, the model predicts that those leaders who are more relationship oriented are more effective in medium situational control and that those who are more task oriented are more effective in high and low control situations (Goethals et al., 2004).

The second approach of relevance to the field of communication is "Transformational Theory" (Burns, 1978). Burns concluded that there were two types of leaders located on opposite sides of a spectrum of leadership. On one side of the spectrum were those leaders who were simply transactional, exchanging one thing for another (e.g., jobs for votes) while on the other side of the spectrum were those leaders who were transformational. The transformational leader offered a transcendent purpose as part of their mission that addressed the higher-order needs of their followers and themselves (Goethals et al., 2004). Higher-order needs included a sense of ethics and ultimately transformed both leader and follower into moral agents.

Finally, the third approach that has relevance here is "Relational Theory" (Komives, Lucas, & McMahon, 2013). The premise of this theory is that the role of relationships cannot be underestimated in the leadership process. Moreover it states that leadership is inherently relational and communal where the interactions that leaders have with individuals and groups has more impact than any speech or persuasive writing could. It is an aspirational model for leaders to develop and support healthy, ethical, inclusive, empowered groups with a sense of purpose that is only achievable together with those they lead.

In addition to an overview of the above leadership theories, it is also useful to note what has been discovered in the areas of leadership study that is most related to the field of communication. According to Modaff et al. (2012), leadership is ultimately a transactional process and that these transactions can only take place through the process of communication. With this in mind, studies in the areas of leadership effectiveness, leadership characteristics, and leadership styles are reviewed in the charts below as these bodies of work best connect to how leaders and followers may use communication skills when participating in the process of leadership. Presenting these areas provides a necessary framework for understanding the importance of the types of leadership training (both implicit and explicit) that take place in the college media context.

Leadership Characteristics and Styles

Current scholarship about leadership characteristics has a wide variety of research lines. Characteristics refers to the attributes and personality qualities that those in leadership possess and demonstrate to others (Clawson, 2011). Leadership characteristic literature ranges from a focus on how leaders think and motivate themselves, to how they work with and inspire those that choose to follow (Bass,

Bernard, & Ruth Bass, 2008). Much of the leadership characteristic literature has measurable communication behaviors e.g., verbal and nonverbal communication strategies, yet to date there has been little to no scholarly attention given to this intersection. Achievement motivation, authenticity, charisma, creativity, ethics, hope, humor, emotional intelligence, leading by example, motivation, optimism, resiliency, risk-taking, and fostering trust are all among the areas where rich communication-focused research could take place (Badaracco & Ellsworth, 1989; Bennis & Biederman, 1997; Bono & Judge, 2003; Conger & Kanungo, 1998; Goleman, 2005; Hardin, 2002; Hartman & Harris, 1992; Masten, 2001; Priest & Swain, 2002; Snyder 2000; Terry, 1993; Winter, 2002). Though the literature on these topics varies in overall focus, ultimately, those in leadership must demonstrate competence in these characteristics in order to achieve success in the groups and organizations within which they work; and communication (with others and with themselves) is a potential vehicle by which they can accomplish this. Moreover, it is the assumption that these characteristics can be developed via the types of leadership training that undergird this chapter.

The literature that explores leadership style aligns with the focus of this chapter as it illuminates what students discover about themselves and what goals, in terms of leadership style development, they make for themselves while working within a college media context. In 1939, social psychologist Kurt Lewin postulated three leadership styles in an infamous study: autocratic, democratic, and laissez-faire. Since then, a whole host of other leadership styles have been categorized and more undoubtedly are to come (Hoyle, 2006). For now though, the most useful styles that have been identified to lead to successful leadership are innovative, reconstructive, shared, strategic, and transformational or transactional. What many of these productive leadership styles have in common is that they support the idea that those leaders who care about the people who follow them achieve better outcomes (Goethals et al., 2004).

Leadership Training and Development

Goethals et al. (2004) write that leadership is "inherently a multilevel phenomenon," which involves individual, groups, and organizations and that leadership development (which includes the specific development strategy of training) is "the act of expanding the capacities of individuals, groups, and organizations" (pp. 840–841). Moreover, they highlight the growing need for organizations to face challenges that are adaptive (situational challenges that are new and must be handled for organizational survival) rather than only technical (reoccurring challenges with go-to solution protocols already established). They argue that as leadership development within organizations continues, strategies that prepare people for more complex forms of leadership would be more and more necessary for those in all roles and levels. They detail seven types of leadership development categories.

Formal Classroom Programs	Classroom seminars that cover basic theories and principles of leadership as well as activities designed to enhance self-awareness and self-understanding.
Outdoor Challenges	A popular development strategy for team-based leadership skills development as well as overcoming risk-taking fears. Includes rope climbing, whitewater rafting, rock climbing & repelling, mock boot camps, and paintball games.
360-Degree Feedback	Known as "multi-layered feedback," that collects the perceptions of peers, subordinates, supervisors, and self.
Executive Coaching	Goal-focused, one-on-one, short term individual sessions often used as a follow-up to 360-degree feedback development strategies.
Mentoring	Formal and sometimes informal development that allows those in leadership to observe and interact with people who are senior to them and with greater experience.
Job Assignments	Specific developmental job assignments that challenge an individual to use leadership approaches that he or she has not used before.
Action Learning	Structured, continuous, and real time processes where organizational members are asked to address an actual problem while also reflecting on their overall performance and learning from a leadership development standpoint.

Along with the above, Velsor, McCauley, and Ruderman (2010) forecast that leadership development efforts in the future would have to take into account strong ethics, technological savvy, and cultural awareness in particular.

Leadership Training and Development in the College Context

Komives, Lucas, and McMahon (2013) are the leading scholars in conceptualizing leadership training and development for college students. They offer their relational leadership model (discussed earlier in this chapter) specifically adapted for the younger and newer-to-leadership training and development audience. Their model calls for the college student to develop best leadership practices in understanding self, understanding others, maintaining integrity, creating communities, productive team & group interaction, and understanding strategic change. Wagner and Ostick (2013) supplement this important work with a series of workshop activities that have a proven track record of working best with the college student audience many of which fall under the Formal Classroom Programs and Outdoor Challenges leadership development categories previously described. Beyond these sources, there lies a great deal of unexplored terrain in the sub-area of college student leadership training and development.

LEADERSHIP TRAINING VOICES FROM COLLEGE MEDIA

In an effort to conceptualize the student leadership training experience in college media, six interviews were conducted for this chapter with those who had direct experience with leadership training in this context. Three interview pairs consisting of a station manager and one of their current or past students participated in an individual interview where they discussed their leadership training experiences while working in a common college media environment. For the sake of anonymity, when referring to each student interviewee and any content from their individual interviews, Student #1, Student #2 and Student #3 will be used. Similarly, the station managers and any content from their individual interviews will be referenced by using Station Manager #1, Station Manager #2 and Station Manager #3. The first interview pair (Station Manager #1, Student #1) consisted of a current professional college radio station manager and his current student station manager. The second interview pair (Station Manager #2, Student #2) was made up of a former professional college radio station general manager and his former graduate assistant. And, the third interview pair (Station Manager # 3, Student, #3) was made up of a current professional college radio station manager and her former news director, a position the student earned during his senior year as an undergraduate.

When asked about their career objectives in media, the students provided a range of responses from working professionally in radio, working in the entertainment industry, and becoming a mass media academic. Anecdotally, both of the now former college media students who took part in the interviews did in fact achieve their career objectives and attributed some of their success to their experiences (inclusive of the implicit and explicit leadership training they received) in college media.

This section of the chapter presents key interview data that sheds light on what leadership development is or was taking place within the college media experience of the interviewees and by extension what may also be occurring within college media contexts more broadly. Of the three interview pairs, two of the supervising professional station managers offered explicit leadership training at some point while working with students. One shared that she used this training as an incentive when attracting talented student workers. All three supervising professional station managers (current and former) reflected that they provided implicit leadership training when needed and perceived these instructive moments to be very helpful for their student employees. From the student or former student's perspective, the explicit *and* implicit leadership training was certainly a draw to working within a college media context along with getting hands-on training in their desired media or academic careers. Student #3 said that when he learned he would receive explicit leadership training once he worked at his college radio station, he found this "very intriguing" and said "it was going to expand my skills set and it would teach me a lot about being a leader." In others words, it influenced his decision to pursue work experience in this context (Student #3, personal communication, October 1, 2014).

There were two specific interview questions that yielded answers particularly relevant to this chapter; this data is categorized by each question below.

WHAT TYPES OF LEADERSHIP TRAINING ARE OFFERED IN YOUR SPECIFIC COLLEGE MEDIA CONTEXT?

The types of training(s) that interview subjects spoke of are listed in the table below.

Type of Leadership Training	Leadership Development Category
Regular one-on-one conversations with advisor or station manager	Executive Coaching
Leadership Theory sections in broadcasting classes	Formal Classroom Program
Leadership role (held by student or former student) in the College Media Context	Job Assignment
Regular advice from former student leaders	Mentoring
Ropes Team Building Training	Outdoor Challenges
Diversity Training	Formal Classroom Program
Intensive Experiential Exercises	Action Learning
Leadership Games & Activities	Formal Classroom Program
Formal 4-Day Leadership Training	Formal Classroom Program
Self, direct report and Station Manager Formal Evaluations	360-degree Feedback

As can be seen in the above chart, the variety of leadership training opportunities offered to students correlates to a number of the leadership development categories established by Goethals et al. (2004), with the formal classroom programs format dominating. What was particularly striking was that many times neither the professional station managers nor their students had a formal name for the type of leadership training they were enacting, but still found these activities influential in creating a positive and productive work culture in their media context. More detail about each of these forms of leadership development is now offered.

Conversations with Advisor or Manager (Executive Coaching)

All three student interviewees spoke about the value of the informal conversations they had with their supervisors. Many of these conversations centered on how they could navigate various leadership dilemmas they faced. The conversations addressed how to handle those students who did not take initiative, how to run meetings with those they supervised, and how to maintain a high standard of professionalism as the academic year progressed. These supervisor conversations served as critical leadership development moments for the students and in reflection were experienced as forms of executive coaching by all. Student #1 said that the informal conversations with her supervisor (Station Manager #1) "help(ed) us more than anything because we were able to discuss leadership among the staff, what we need to do … and be more organized (Student #1, personal communication, October 10, 2014). Student #3 said that he and his supervising Station Manager frequently "sat down a lot and talked about things I could do to help my other peers to get on board with the vision" (Student #3, personal communication, October 1, 2014).

Leadership Theory in Class (Formal Classroom Programs)

Station Manager #3 was able to regularly work with her student employees in a class that she organizes throughout the academic year and for which she serves as professor. In this class, she supplements the hands-on training that occurred at the on-campus radio station with assigned leadership development readings and theory. More specifically, she said "there is a leadership packet that I made with goals for each week that needed to be accomplished. We are always trying to make leaders out of students that have a lot of potential" (Station Manager #3, personal communication, October 10, 2014). Station Manager #1 also worked with some of his student employees in a broadcasting class. He said "In that class, we talk about theories about managers and management" (Station Manager #1, personal communication, October 6, 2014). Station Manager #1 indicated that covering these theories in class laid the groundwork for more applied conversations in the college media context and generated several "A-Ha" moments for the students.

Formal Leadership Roles in College Media (Job Assignment)

Two of the three students that were interviewed actually held formal leadership roles (Student Station Manager and News Director) while working within their college media contexts and were not volunteers only. They each were student station managers and both found having this role valuable in terms of leadership development. Student #3 said:

> I wanted to see what I could do to make this the best possible experience and to be a good leader. It's kind of intimidating to be so close in age to those I was leading. So, it was about how I can go about effectively leading this group of people who either, I am the same age as, or some of them could potentially be older than me. (Student #3, personal communication, October 1, 2014)

Advice from Former Student Leaders (Mentoring)

Station Manager #3 required Student #3 to do two hours a week of peer mentoring in which technical skills were imparted. Station Manager #2 also had Student #2 have informal leadership development conversations that supported the newer students as they took up a leadership role on campus and in many ways were representing the university. Student #1 also sought out mentoring from former radio station student employees. She said her conversations with a former student station manager were particularly helpful and said "I'd go to him and be like … what advice do you have for me? How do you do what you do?" (Student #1, personal communication, October 10, 2014).

Team Building Training (Outdoor Challenges)

Station Manager #2 and Student #2 spoke about an outdoor ropes course they participated in. Both interviewees noticed a substantially positive impact on how this experience shaped the way their college radio station staff worked with each other and moved the station forward to success. Station Manager #2 said "in my last few years, we would bring the students into training. We did this at least two or three times … a specific off-campus ropes training with the students. And it was absolutely wonderful and you could really see the students learning through these exercises" (Station Manager #2, personal communication, October 7, 2014). Student #2 said:

> This was the first time I ever remember anything being branded as far as leadership. The entire premise behind having a responsibility as a director, you were obviously a leader. But, you had a very big role to fill and you had a supporting cast of folks that had to support what you did for each area. (Student #2, personal communication, October 12, 2014)

Diversity Training, Games and Activities (Formal Classroom Program)

Two of the three station managers incorporated a wide variety of formal classroom programming. Station Manager #2 spoke about the increase in formal training that his university offered during his time with the FM radio station. While he could not recall the specifics of the trainings, he talked with a former colleague and was reminded of several specific trainings that took place. These included a review of Janet and Milton Bennett's "Intercultural Communication Model," a modified Tavistock training on personal cultural understanding and the review of an interesting human relations model called "The New White Person" which was about building awareness of minority and majority status. Station Manager #2 also remembered several structured experiential activities that he and his student worker team played that had a leadership development focus. Station Manager #3 asked her student employees to play a phone etiquette game which she saw as an important part of leadership development and said "And to you and me, it may seem like second nature, but the students who didn't know how to pick up a phone and formally call someone for a phone interview … that told us a lot" (Station Manager #3, personal communication, October 10, 2014.)

Experiential Exercises (Action Learning)

Station Manager #2 spoke of several successful experiential exercises that held meaning for the student participants. One of them focused on breaking cultural stereotypes, while another dealt with the power of varying perceptions among members of a team. He saw these exercises as an integral part of the leadership development that took place for the students that worked at his radio station. Of these types of activities, he said "It would remind people about the roles they would have. I would see that with the radio station students. For a couple of them, the light bulb would go on" (Station Manager #2, personal communication, October 7, 2014). In terms of the impact of these trainings, he said "I saw people being a little more focused, people getting along a little bit better, maybe even *more* than a little bit better. We became more productive as a team. Just the fact that people treated each other better" (Station Manager #2, person communication, October 7, 2014).

Leadership Training (Formal Classroom Programs)

Two of the three station managers offered what they considered to be formal leadership training that was regular and expected. Station Manager #1 held a week-long leadership training which all employees were expected to attend. The focus of these workshops varied from personal development (communication styles) to team development (managing differences). Of one workshop she said "We bring in alumni speakers, those types of things, to help motivate them" (Station Manager #3, personal communication, October 10, 2014). Her student employee remembered the leadership development that these workshops had on him as he reflected his sense of responsibility once he moved off campus. He said:

> We were going beyond the walls of the university. That just shows that when we are in that community setting, we have a role to fill. We have to be that sort of leader. We need to be able to represent the station well. (Student #3, personal communication, October 1, 2014)

Student #2 reflected on the formal leadership training that she and other students experienced when working for Station Manager #2. What she remembered captured the value of these formal leadership development workshops. "There were a series of external workshops that we did on leadership training and formal meetings. Leadership was there. It was an obvious role they [the students] had to play. And that expectation was there from day one" (Student #2, personal communication, October 12, 2014).

Evaluations (360-degree Feedback)

Only one station manager (Station Manager #3) reflected on a kind of 360-degree evaluation that occurred at her radio station. She required feedback evaluations for all student employees of themselves, by a peer and from herself as supervisor. These evaluations were all in an effort to help the students develop goals in their roles as leaders. When the interviewer (the author of this chapter) introduced the idea that this was a form of 360-degree evaluation, she agreed and said "Yes … it has components of 360 degree evaluation to a certain extent" (Station Manager #3, personal communication, October 10, 2015).

WHAT DO YOU THINK THE IMPACT IS OR WAS OF THE LEADERSHIP TRAINING(S) AND YOUR EXPERIENCE IN THE COLLEGE MEDIA CONTEXT?

Below is a selection of particularly informative responses to this second question:

- Having that leadership training, being given these strict expectations, and being told how to represent this radio station well, and how to be this leader in this community—it helped me to be a stronger leader in all of my other roles. (Student #3, personal communication, October 1, 2014)
- I feel like my leadership style was influenced by the charge that was given to me and I have kind of transferred it over into my life. (Student, #3, personal communication, October 1, 2014)

- … from my viewpoint is that you want to know that it prepares me—and I feel like I am 'cause I feel like what I went through here is a real life situation. I didn't get this really long experience where I was told how to do all these different things. I was thrown into it. And I learned how to work with people and I learned when I needed to take questions, and I learned what I needed to do. And I think that is real world. (Student # 1, personal communication, October 10, 2014)
- I've learned so much about what I need to do and about the fact that if I go out into the real world, I can say I have led a lot of people. (Student #1, personal communication, October 10, 2014)
- … I'm depended on by so many people here. It's like real life. (Student #1, personal communication, October 10, 2014)
- I noticed students became very good at problem-solving and also supporting one another. (Student #2, personal communication, October 12, 2014)

From this selection of statements it is clear that the variety of leadership development opportunities (both implicit and explicit) was useful, had meaning and perhaps most importantly allowed students to transition from a college campus work environment to a professional work environment with a greater set of leadership skills and capabilities.

Finally, Student #2 identified with an analogy during her interview (offered by the interviewer) that crystalized for her the unique way that the college media work environment brings together normally separate and diverse groups of people. She said that students find the college media context, "… unique in the college world because unusually, they [students] don't get to interact like that, unless they are in a specific group" (Student #2, personal communication, October 12, 2014). When hearing this, the interviewer offered the metaphor of the college media context being like *The Breakfast Club*: Meaning that the unique setting of college media brings together a diverse set of students who share a common experience that molds them—at least some—for their professional lives.

The motion picture *The Breakfast* Club, written and directed by John Hughes, presents a diverse group of students (prom queen, star student athlete, troubled teen, social misfit and a nerd) who share a Saturday afternoon detention session and bond in unexpected ways. The notion of unexpected bonding seems to also be found in the context of college media itself and can potentially be attributed to the leadership development that takes place within it. In other words, it could be argued that the diversity that the college media context offered the interviewees more than likely mirrored the workforce that many of them entered as media professionals and academics. And that with the impact of the leadership development they each had, they were more capable leaders for the world that awaited.

CONCLUSIONS AND RECOMMENDATIONS

When looking at the above interview data and reviewed leadership development theory, it becomes clear that *leadership training* plays a role in establishing the unique nature of the student experience in college media. What seems to characterize this impact best is that the students learned about leadership both within implicit and explicit leadership training modalities. Moreover, it can be argued that the professors, advisors, and station managers who were interviewed represent a desirable direction in the management of college media: offering student leadership training as an explicit benefit to working in this context, a benefit that students value. Finally, it is clear that examining the process of leadership (and all that it involves) within the mass communication context is virtually non-existent and that this chapter offers one of the few important focuses on this nexus. Leadership influences a variety of important mass communication questions, contexts, and research directions. Exploring how this is so, is an important area for further research. The very act of asking these types of new questions within mass communication is an act of leadership in itself, and could logically influence other equally less studied fields to date (i.e., STEM fields, Social Work, Architecture) about how implicit and explicit leadership development can benefit students and better prepare them for professional life.

REFERENCES

Badaracco, J. L., Jr., & Ellsworth, R. R. (1989). *Leadership and the quest for integrity.* Boston, MA: Harvard Business School Press.

Barrett, D. J. (2011). *Leadership communication* (3rd ed.). New York, NY: McGraw Hill Irwin.

Bass, B. M., & Ruth, B. (2008). *The Bass handbook of leadership: Theory, research, and managerial applications* (4th ed., Free Press hardcover ed.). New York, NY: Free Press.

Bennis, W., & Biederman, P. W. (1997). *Organizing genius: The secrets of creative collaboration.* Reading, MA: Addison-Wesley.

Bono, J. E., & Judge, T. A. (2003). Self-concordance at work: Toward understanding the motivational effects of transformational leaders. *Academy of Management, 46*(5), 554–571.

Burns, J. M. (1978). *Leadership.* New York, NY: Harper Collins.

Clawson, J. (2011). *Level three leadership: Getting below the surface* (5th ed.). Upper Saddle River, NJ: Pearson Education.

Conger, J. A., & Kanungo, R. N. (1998). *Charismatic leadership in organizations.* Thousand Oaks, CA: Sage Publications.

Fainhurst, G. T., & Grant, D. (2010). The social construction of leadership: A sailing guide. *Management Communication Quarterly, 56*(3), 215–239.

Friesen, G., & Meyer, A. (Producers) & Hughes, J. (Director) (1985). *The breakfast club* [Motion picture]. United States. Universal Studios.

Gamble, T. K., & Gamble, M. W. (2013). *Leading with communication: A practical approach to leadership communication.* Thousand Oaks, CA: Sage Publications.
Gardner, W. L. (2003). Perceptions of leader charisma, effectiveness, and integrity: Effects of exemplification, delivery, and ethical reputation. *Management Communication Quarterly, 16*, 502–527.
Goethals, G. R., Burns, J. M., & Sorenson, G. J. (2004). *Encyclopedia of leadership.* Thousand Oaks, CA: Sage Publications.
Goleman, D. (2005). *Emotional intelligence: Why it can matter more than IQ.* New York, NY: Bantam Books.
Hardin, R. (2002). *Trustworthiness.* New York, NY: Sage Publications.
Hartman, S. J., & Harris, O. J. (1992). The role of parental influence in leadership. *The Journal of Social Psychology, 13*(2), 153–167.
Hershey, P., & Blanchard, K. (1969). *Management of organizational behavior.* Upper Saddle River, NJ: Prentice Hall.
Hoyle, J. R. (2006). Leadership styles. In *Encyclopedia of educational leadership and administration.* Retrieved from http://dx.doi.org/10.4135/9781412939584
Jackson, B., & Parry, K. (2011). *A very short, fairly interesting and reasonably cheap book about studying about leadership.* Thousand Oaks, CA: Sage Publications.
Komives, S. R., Lucas, N., & McMahon, T. R. (2013). *Exploring leadership for college students who want to make a difference* (3rd ed.). San Francisco, CA: Jossey-Bass.
Masten, A. S. (2001). Ordinary magic: Resilience process in development. *American Psychologist, 56*, 227–239.
Miller, K. (2015). *Organizational communication* (5th ed.). Stamford, CT: Cengage Learning.
Modaff, D. P., Butler, J. A., & Dewine, S. (2012). *Organizational communication: Foundations, challenges, and misunderstandings* (3rd ed.). Glenview, IL: Pearson Education.
Norhia, N., & Khurana, R. (2010). *Handbook of leadership theory and practice: An HBS centennial colloquium on advancing leadership.* Boston, MA: Harvard Business Press.
Northouse, P. G. (2015). *Leadership: Theory and practice* (7th ed.). Thousand Oaks, CA: Sage Publications.
Priest, R. F., & Swain, J. E. (2002). Humor and its implications for leadership effectiveness. *Humor, 15*(2), 169–190.
Snyder, C. R. (Ed.). (2000). *Handbook of high hope: Theory, measurement, and applications.* San Diego, CA: Academic Press.
Terry, R. W. (1993). *Authentic leadership: Courage in action.* San Francisco, CA: Jossey-Bass.
Velsor, E. V., McCauley, C. D., & Ruderman, M. N. (Eds.). (2010). *The center for creative leadership handbook of leadership development* (3rd ed.). San Francisco, CA: John Wiley & Sons.
Vires, E. R., Bakker-Pierpr, A., & Aastenveld, W. (2010). Leadership = communication? The relations of leaders' communication styles with leadership styles, knowledge, sharing and learning outcomes. *Journal of Business and Psychology, 25*(3), 367–380.
Wagner, J., & Ostick, D. T. (2013). *Exploring leadership for college students who want to make a difference.* San Francisco, CA: Jossey-Bass.
Winter, D. G. (2002). The motivational dimensions of leadership: Power, achievement, and affiliation. In R. Riggio, S. Murphy, & F. J. Pirozzolo (Eds.), *Multiple intelligences and leadership* (pp. 119–138). Mahwah, NJ: Erlbaum.

SECTION TWO

Description AND Narrative

CHAPTER FOUR

Why College Radio?

LEN O'KELLY

Over the last twenty-eight years since I first set foot into a college radio station I've had numerous opportunities to share stories about my time on the air at the "low end" of the dial. In a broadcasting career that spanned several states and two continents, it was commonplace to find others who had similar experiences. Even outside the industry, though, it was relatively easy to find someone who either had worked at their college station or was a close friend with someone who did (and, as such, was invited down into the basement of a dorm to sit in on a show). For me, college radio was solely responsible for getting my career started. Had I not wandered into the studios of WLRA-FM at Lewis University in Romeoville, Illinois, one day in 1987, I likely never would have gone into the field. It was there that I learned the nuts and bolts of radio but also made lifelong friends and business contacts that started the wheels of a career in motion.

My story is by no means unique. Mark Manuel, longtime morning radio host in Davenport, Iowa, began his career as a college radio disc jockey at Iowa State University almost thirty years ago. What made the experience valuable to him was meeting his student radio station manager, Lou Sipolt, who became a mentor.

> He was so passionate about radio and saw that I was developing the same feelings, so he took me under his wing and, after my freshman year of pulling the overnight shift once a week, helped me land my first commercial radio job at KCCQ-FM/KASI-AM in Ames, Iowa. (M. Manuel, personal communication, December 2014)

Peter Kreten, who rose through the ranks from student member to full-time general manager of WXAV-FM at Saint Xavier University in Chicago, IL, points to college radio's ability to provide a laboratory setting for students.

> If you are a biology major or a chemistry major, you have to attend labs where you experience the laws of the universe first hand. It's the same with college radio. The only way you become "good" at broadcasting is by doing it. And that is what college radio is for. It is where you find your feet. (P. Kreten, personal communication, December 2014)

Even a legendary air talent such as Bob Hale, who as a young Dubuque disk jockey hosted the last performance given by Buddy Holly on the "Winter Dance Party" in 1959 and later became one of the original stars of WLS/Chicago's rock and roll format, wouldn't have gotten his start in radio had it not been for the opportunities provided to him by the equally-legendary WHA at the University of Wisconsin-Madison.

Both the times and the technology have changed dramatically since the late 1980s. The Internet now makes it possible for audiences to cross borders in ways never thought possible with traditional radio signals. The computing power and storage space of laptops make it possible for a student to possess a music collection the size of what most of our radio stations had available—and the means to play it in a prescribed sequence. The basics of podcasting are within reach of almost any college student. Theoretically, any and every student could "broadcast themselves," to steal the phrase from YouTube, leaving little or no need for an institutionally supervised and funded facility.

Yet tens of thousands of students at thousands of student radio stations choose not to go it alone. Instead, college students in the new millennium are still fascinated by traditional college radio. They still gather together in the basement of a campus building, play music or engage in conversation, and share in the same experience that students have been participating in for almost one hundred years. In exchange, they adhere to commitments and follow rules laid down by faculty advisors and university administrators with respect to content. What motivates, them, however, is the value of the experience in terms of a future career—and they cannot get that advantage sitting by themselves in a dorm room. The ability to network with professionals before graduation gives students entering a difficult job market an advantage. For many students who come from high schools with extensive media programs—something much less common even ten years ago—the organized structure of advised student media can even help with the sometimes difficult transition from high school to college life in general.

It is important to consider the value of student media as a supplement to the education that students receive while enrolled. I had the opportunity to spend time with a recent student of mine who has graduated and begun his first full-time position in radio. The student shared that, in the past year he has realized that the skills that benefitted him most were not learned in the classroom but rather in the student radio station. His experience has more in common with what students have experienced for the last fifty years.

A BRIEF HISTORY OF BROADCASTING PEDAGOGY

As far back as 1966 Keith Brooks argued that critics of radio and television courses thought that they were "unsubstantial" and "at best, extracurricular" (Brooks, 1966); his suggestion was to add discussion of communication theories and models as well as in-depth conversation about the role of broadcasting and the public interest. While these topics are important, the changes failed to address the lack of practical training. Most students were still not seeing this sort of experience until after graduation (Balon, 1978). Balon cited professional broadcasters who lamented having to "spend six months" bringing new graduates to a level of understanding they needed to be successful broadcasters. His was not the only study to find this to be the case: Research by James S. Taylor in 1974 surveyed 147 radio and television station managers as well as 51 college and university broadcasting department chairs to see how effective they felt the education being provided to students was. Taylor found that 52 percent of radio station managers and 80 percent of television station managers found a college education to be "useful" in obtaining a job in broadcasting. Interestingly, only 84 percent of the educators felt the same way (Taylor, 1976).

What the student radio station offers as a working laboratory is invaluable. Part of the perceived inadequacy of the classroom experience is its artificiality: Are students really being prepared for what they can expect in the workplace under real-life conditions? Programs have been attempted at schools such as Cal State Northridge that took students from having no experience to immersing them into newsroom-like settings and removed the classroom component, in a sense (Marks, 1999). An additional point to consider is how well students are critiqued for the hands-on work that they do. Students are unique as compared to established broadcasters in that they are typically more receptive to being coached (Kozak, 1997) and effective instructors spend the time to critique student work. Traditional studies of the effectiveness of college's role in career preparedness tend to focus on field of study and grade point average (Pascarella & Terenzini, 1991) and overlook the student media experience entirely. What do those students who have made the transition from student media to professional broadcast media say about the value of being a part of the organized media program?

THE STUDENTS SPEAK UP

The question of "why college radio" provided the opportunity to converse with a number of former students and college media participants who have gone on to work in the field of broadcasting. While the stories of their experiences were different, some common themes began to emerge. The first theme was "perceived educational value." One went so far as to say that her experience in student media "helped me more than the classes that I have taken." Another admitted sheepishly that he had devoted more time and effort to the student media than to his work in the classroom, but also admitted that he did not regret that decision. "I owe my career to that experience" (Anonymous Grand Valley State student, personal correspondence, April 2014).

Networking is particularly valuable to today's college student. We spend time in the classroom stressing the importance of making connections. The connections that can form in the student radio station can be important at all phases of a student's career. I often share with my students, for example, that I am still in contact with about a half-dozen or so people with whom I was on the air in the 1980s. Some of those connections are limited to the occasional Facebook message, while others take their shape in the form of being a part of family milestones such as births and deaths. But there is also a business component. Freelance opportunities and business leads can come from these college radio relationships, and students who are savvy enough to make the most of their time at the station can reap the benefits of these contacts. One former student specifically mentioned that the connections he made through students involved with media (and with professors who worked with those students) were invaluable in helping him to land his first broadcasting job. He is still in contact with many of those students as he plans his transition from entry-level job to his next position in the field.

Networking with other students is important, but those involved in student media often have the opportunity to network with a professor or other seasoned professional. Students often pointed to the advice and mentoring that they received during their time at the station as being particularly valuable. One said that he "went into the station alone, but came out with both friends and a mentor" (Anonymous student, personal correspondence, April 2014). Mackenzie Thaden, now a producer at WZZM-TV in Grand Rapids, Michigan, indicated that the opportunity to talk with a trusted adult about difficult career decisions, such as whether or not to take a first job away from her family and her boyfriend, was of particular comfort. "He gave me advice and was a sounding board when I needed it" (M. Thaden, personal correspondence, April 2014).

Professional skill-building, for lack of a better phrase, is something that classroom instructors struggle to teach. It can be challenging to get students to accept the artificial construct of the classroom and understand how time management and collaboration will be of benefit "in the real world." Student media, however, demonstrates this to students on a continuous basis. One student said that working in student media taught her the importance of deadlines and presented the challenge of having to come up with compelling content on a more frequent and regular basis. Perhaps most notably, another pointed to the value of student collaboration through group work on productions as likely being valuable. Yet another was more pragmatic: the student station was "a place for her to make mistakes" before "having to do it for real" (Anonymous students, personal correspondence, April 2014). Mackenzie Thaden realized that student, media and college in general helped to prepare her for work in ways she did not expect.

> In the same way you have to learn the way a professor grades or a group dynamic functions, I had to learn exactly where I fit in with the other people, the newscast, and the newsroom. But for every new job I will have to do that. (M. Thaden, personal correspondence, April 2014)

While student media allows the comfort of community, it also can provide a student with the ability to become comfortable with their own independence. Dan Spadafora, a former student who worked professionally before graduating and who is currently program director of WHTC-AM in Holland, Michigan, summed it up this way:

> Once you are on the job it's more about learning how "real" radio or television operates compared to what you are exposed to [at school]. You are more on your own once on the job. No one is going to hold your hand and show you what to do. There will obviously be training to do the job but in every job I have had in broadcasting, people are stretched so thin that most of the "training" is done by yourself. There needs to be a way to prepare students for this self-reliance they will need. (D. Spadafora, personal correspondence, April 2014)

Mackenzie Thaden had done both student radio and student television as an undergraduate. She was very insightful about what she learned from her student media experiences: They taught her more than just how to operate a radio station.

> I did not put in the time that I needed in order to make my show good, and depended too much on a co-host that didn't show up a lot. The experience taught me a lot about myself and my choices … almost more so than actually about radio. (M. Thaden, personal correspondence, April 2014)

For many of the students interviewed, the ability to accept a greater role in the radio station had more value than just a title. The opportunity to lead and/or train other students was cited as being valuable or worthwhile. Dan Spadafora specifically pointed to the opportunity that he had as the general manager of his student station as "showing me that I could, in fact, lead the troops. I never thought that I could" (D. Spadafora, personal correspondence, April 2014).

SHOULD THE SCHOOL KEEP ITS STATION?

College and universities that jettison their student radio and television stations may do so at their peril. Student involvement in college is a key determining factor of student success, satisfaction, and persistence (Astin, 1984) and students who become involved in organizations outside of the classroom while at school develop increasing pride in their institution (Abrahamowicz, 1988). But while Shertzer et al. (2005) cite Howe and Strauss (2000) among others in suggesting that Millennial students have "been involved with more scheduled activities" than other students before them and therefore may "not need to be persuaded to get involved," the opposite may in fact be true (pp. 85–86). Without providing opportunities for students to join an organization, and publicizing said opportunities, chronically over-supervised students may not seek the opportunities on their own. A well-run student media organization can therefore serve two purposes: first, it can provide a club for students to join, and in which eventually to take ownership and pride, and second, it can help make the greater campus community aware of other clubs and organizations in which students both inside and outside of the radio station can take a participatory role. It is also not a huge leap to realize that an alumnus/alumna who is more happy about their college experience—one who feels more pride—is more likely to eventually become a financial supporter of their alma mater (Anik et al., 2011).

If there was an area of student media experience about which students tended to be more critical, it was the topic of how those media outlets are managed at some institutions. Grand Valley State University, where I teach, puts an emphasis on its media being "student-run:" The media outlets are not formally affiliated with the academic department but have a faculty advisor to assist students. Students who took part in these types of student media cited this as more of a drawback than a benefit. Dan Spadafora summed it up:

> More structure would help. I firmly believe the groups should be student-run, but even I can admit sometimes I felt lost when I led [the station] ... it could be helpful if some decisions and structuring was done by the advisor. Students are students and this should be an opportunity to learn how to lead and work in the career they want in a safe environment, but [students] making all the decisions could hurt the group in the long run. (D. Spadafora, personal correspondence, April 2014)

Perhaps surprisingly, several other students spoke favorably about increased structure, preferring that approach to the free-for-all often associated with student radio (and certainly something that would be absent if students were left to their own devices). Organization and professionalism was a recurring theme: One student thought that those participating in student media should get more feedback from professors about their work. Another cited "not getting help from other students" as a negative feature. A third specifically mentioned the lack of funding as a drawback as it caused lags in technology. She went as far as to recommend that there should be more courses that require participation in student media "because perhaps that way there would be more funding and resources available for the station" (Anonymous student, personal correspondence, April 2014).

An anonymous student who spoke of the value of his broadcasting education but did not participate in student media stated that he saw the value in it but could not find time between internships, coursework, and working to pay tuition. Yet another agreed that while her student media program had shortcomings, "I think that there were a lot of opportunities for students to really do well, it's just a matter of whether or not students actually want to take on those opportunities" (Anonymous student, personal correspondence, April 2014).

THE ROLE OF STUDENT MEDIA

Is participation in student media that important? I asked graduates which part of their education was "most valuable" to them, but most were wary to point to one component of their experience as being the sole agent to successful career preparation. When asked to choose between student media, formal professional internships, or classroom work, they were more likely to speak to the value of all three methods. Alana Holland, morning news anchor at WNEM-TV in Saginaw, MI, said that while her internships were most beneficial, "I would not have felt as comfortable and confident in my internship experience without the help of my course materials," and added that "the students who do all three are going to have the best experience overall" (A. Holland, personal correspondence, April 2014). Allison Newman, a senior and station manager of Grand Valley's student television station, looked past coursework, stating that on-the-job training and extra-curricular work had been most beneficial: "You can only learn so much from a book. It's when you get out into the field and use equipment that you actually gain the knowledge you need" (A. Newman, personal correspondence, April 2014). One suggested a reason for the effectiveness of student media that is not often considered: the ability to learn skills and equipment at one's own pace rather than one prescribed by an instructor. Among more seasoned professionals the responses differed. Mackenzie Thaden simply said:

> You need it all. Student media allowed me to create. Without the class work, I wouldn't be able to do what I am doing now, and the internships allowed me to use my skills and to get a leg up on the competition. (M. Thaden, personal correspondence, April 2014)

Kayla Fortney, who worked as a producer at both WILX-TV in Lansing, MI and WZZM-TV in Grand Rapids, said that her hands-on experience in class and in student radio as well as in a news writing course "taught me the most and have given me my biggest advantage in my jobs" (K. Fortney, personal correspondence, April 2014).

The demise of terrestrial radio has been written about since the advent of television. The arrival of every new bit of technology—from the television to the cassette player to the cellphone to the iPod to the Internet—has been heralded as the death knell for standard broadcasting. But with a billion radio sets still in use in the United States alone, it doesn't appear that the medium is fading away in the immediate future. It would be more likely to predict the fall of radio if the youngest users—high school and college students—were avoiding it in droves. But they are not. Research we conducted at Grand Valley State in 2011—at a time when deciding whether to pursue a low-power FM signal—showed that 85 percent of our students still used traditional radio, and that FM radio was the second most cited source for learning about new music after "recommendations from friends" (O'Kelly, 2011).

Not only are these students using radio, they want to create it themselves. In 2014, over four hundred high school and college radio stations took part in the fourth annual College Radio Day. Originally conceived in 2011 to call attention to student radio in the United States, the event has become World College Radio Day, attracting stations from across the globe. These students come together for this event in the same way that they come together each day in student radio stations instead of hiding in their dormitories alone—the same way that generations before them came together to work on a common project. College radio in the new millennium is still a vital, valuable resource for both education and enrichment.

REFERENCES

Abrahamowicz, D. (1988). College involvement, perceptions, and satisfaction: A study of membership in student organizations. *Journal of College Student Development, 29*, 233–238.

Anik, L., Aknin, L. B., Norton, M., & Dunn, E. W. (2011). Feeling good about giving: the benefits and costs of self-interested charitable behavior. In Oppenheimer, D. M. and Olivola, C. Y. (Eds.), *The Science of Giving*. New York: Taylor and Francis.

Astin, A. W. (1984). Student involvement: A developmental theory for higher education. *Journal of College Student Development, 25*, 297–308.

Balon, R. E. (1978). Into the industry: The student as broadcasting consultant. *Communication Education, 27*(1), 25–39.

Brooks, K. (1966). A framework for broadcasting study. *NASSP Bulletin, 50*(312), 168–172.

Howe, N., & Strauss, W. (2000). *Millennials rising*. New York, NY: Vintage.

Kozak, L. (1997). Dynamics of successful coaching of student talent in broadcasting: An elusive art in the news lab. *Journalism & Mass Communication Educator, 52*(2), 43.

Marks, R. (1999). Beyond the anchor: Students and broadcasting opportunities. *Journalism & Mass Communication Educator, 54*(1), 79.

O'Kelly, L. (2011). A survey of the music listening habits of college students, and the role of student radio in those habits (Unpublished raw data).

Pascarella, E. T., & Terenzini, P. T. (1991). *How college affects students*. San Francisco, CA: Jossey-Bass.

Shertzer, J., Wall, V., Frandsen, A., & Guo, Y. (2005). Four dimensions of student leadership: What predicts students' attitudes towards leadership development? *College Student Affairs Journal, 25*(1), 85–107.

Taylor, J. S. (1976). Broadcast education as career preparation: How good is it? *Central States Speech Journal, 27*(1), 59–69.

CHAPTER FIVE

The Determinative Affect ON Broadcast Development AND Career Outcomes OF Parental Influence AND Changing Peer Group Dynamics

A Goffman Case Study Inside Student-Run College Radio

LEO J. FAHEY

Radios were always around. The family radio was in the kitchen of my Bay Ridge, Brooklyn, home. WOR's personality driven talk, information and middle-of-the-road music was heard every morning. There was also a faux-leather covered Philco portable my family took everywhere. I have a clear memory of listening to the *Lone Ranger* over it. My folks put a clock radio next to my bed and they gave me a transistor for my eleventh birthday. I listened to *Dickey Doo and the Don'ts, Little Eva, Gino and Gina, The Everly Brothers, The Chordettes, The Tokens, The Tremeloes,* and all the other *hit makers* over WMCA and WABC whenever I was allowed to have a radio on. Then I discovered DX'ing: I would stay up well past bedtime searching the skip on the clock radio pulling-in mostly Chicago stations. I felt whatever I was to do in life should be done in radio. But, as an obedient only child I had to comply with my father's demand for high academic achievement, acceptance into an academic collegiate program and graduation into being an officer, a gentleman and a scholar, a scholar-warrior.

This study examines the push and pull of parental power and peer group influence over an individual's career decisions and outcomes.

GROUNDING

Parental influence on their children's career choices is, according to Clutter (2010), consistently important. He cites Ferry (2006) as having found that while schools, peers, and the child's community impact young adults' career choices, parents' expectations and perceptions of "vocational fit" shape the child's decisions. Emphasizing the point, Clutter states that a Creamer and Laughlin study (2005) concluded that parental influence was so strong as to override the influence of teachers, faculty, and career counselors (Clutter, 2010, p. 2).

Bergen (2006) cites Mitchell and Krumboltz (1990) as arguing that parents' active support or discouragement of certain behaviors can encourage or discourage career development directions. Further, Bergen states: "Over involvement or dysfunctional relationships between parents and their children may inhibit the individuation process and contribute to young individuals' indecisiveness due to conflict with their own needs and wants and those of their family" (Bergen, 2006, p. 10).

Parental expectations definitely shaped career decision-making here. Family dysfunction is not argued, but "over involvement" of a father in a son's decisions certainly is. My father needed to live through me to repair his deepest regret. He suffered from asthma from youth. The only time he was free from it was in the U.S. Army during the Second World War. He was a staff sergeant in the Quartermaster Corps who was adept at playing the system to enjoy his most favorite activity, reading great literature. He had not gone to college and the draft had precluded it for the duration. However, as a means of being in the right place, he maneuvered to take the test for admission to the U.S. Military Academy, a number of seats being available for active duty soldiers. He passed the test but for reasons unexplained he chose not to go.

Yet, he always felt not quite right with where his civilian life had taken him after the war. And he continued to suffer greatly from asthma, which eventually cost him his life. He kept saying he terribly regretted rejecting West Point seeing, he thought, life as a military scholar would have made him happier and, most important of all, healthy. He ached for the health those years in the Army brought. That longing was the basis of the deepest of regrets, and the basis of his near maniacal demand I become a scholar-warrior. If he couldn't be personally happy and healthy then, he could best approximate it by living through me.

The struggle between my father's need to live through me and my need to live my own life see-sawed with the changing relationships of the peer groups in the student-run radio organizations through which I labored. A buffer between my father's demands and my needs was created when I was fully a part of the group allowing my radio career decisions to prevail. However, when I was outside the peer group, no such buffer was available and, consequently, my father's dominance overpowered almost all of my career decisions.

Erving Goffman (1959) detailed a method describing social convergences and divergences of people within groups. This provides consistent identification of inclusion within, and separation from, the societies of fellow students of the three student-run college radio stations, WLIU, WNYT and WNYU, through which I sought to develop a radio broadcast career. This study employs three concepts from Goffman—Performance, The Definition of the Situation, and Team.

Performance is the self-image an individual, intentionally and unintentionally, conveys to others. It includes "… all activity of an individual which occurs during a period marked by his continuous presence before a particular set of observers and which has some influence on the observers" (Goffman, 1959, p. 22). Performance's expressive actions, in Goffman's words, define the situation. The Definition of the Situation is the projected behaviors of an individual into a group of other individuals and the behavioral expectations of the individual held by others in the group:

> When an individual enters the presence of others, they commonly seek to acquire information about him … Information about the individual … [enables] others to know in advance what he will expect of them and what they may expect of him. (Goffman, 1959, p. 1)

Defining the Situation is reciprocal between an individual and the others within a group. As Goffman points out,

> When we allow that the individual projects a definition of the situation when he appears before others, we must also see that the others, however passive their role may seem to be, will themselves effectively project a definition of the situation by virtue of their response to the individual and by virtue of any lines of action they initiate to him. (Goffman, 1959, p. 9)

> When reciprocality demonstrates a common definition of the situation within a group, that is, when everyone in a group shares the same range of projected and expected self-image behaviors, individuals will collectivize the definition of the situation and will cooperate with each other to maintain it as the group stages its established patterns of action, [or "routines" as Goffman terms them]. (Goffman, 1959, p. 15)

The commonality of held and projected self-image, the definition of the situation, and consequent behaviors will bring those in communion into an interdependent group that Goffman calls a "performance team" or just a "team" (Goffman, 1959, p. 79). A team projects its collective definition of the situation and consequent behaviors to themselves and to other teams within an organization as well as to all others outside the organization (to "audiences" in Goffman's terms) maintaining the agreed upon self-image and behaviors during the accomplishment of individual and collective, social and work obligations.

The Goffman concepts selected: Performance, The Definition of the Situation, and Team are used here to describe changing relationships with college radio colleagues and subsequent impact on career outcomes these changes can cause.

NARRATIVE

In the early spring of 1969, the radio broadcasting bug bit hard. My cousin Mary Ann is to be thanked for re-igniting the passion. She then worked at WFIL, the Top Forty AM radio station in Philadelphia, PA. She told legions of great stories of the DJ's. She arranged visits. And by late August, I pledged to myself I would find ways into this wonderful work.

Long Island University, Brooklyn, the college I was attending as a Physical Education major, had yet to develop one in Media. But to my incredible good fortune I discovered within the first couple of weeks of the Fall '69 term that a closed-circuit radio station was being constructed. The small student group responsible petitioned throughout the college for additional help. I rushed to get deeply involved!

We were amazingly similar: purposes, interests, aesthetics were pretty much the same; maturity ranged within a year of each other; each had largely the same collegiate experiences. From the similarity emerged a singular definition of the situation: we projected ourselves to be serious, enthusiastic, progressive radio broadcast amateurs on the threshold of turning professional.

We had no radio broadcast experience and less than minimum understanding of broadcast electronics. Steve Lobel, a senior and the group leader, knew how the equipment fit together since he worked with suppliers when writing the proposal. I and others worked closely with him on station construction.

There were no arguments about who was doing what in studio construction. Steve showed us the floor plans of the two attached rooms of the third floor low-rise on-campus office building the Administration had given us for the studios. The equipment had been laid out on the floors. We selected what we wanted to do according to interest. Folks came and went depending on class schedules. When something was to be left to be done later, it was left or it was picked-up by someone else coming in from class.

The studios came together. The Gates audio console was affixed to the main studio table. The two turntables and their amplifiers went into their respective pre-built floor cabinets and were set in front of the console on each side of the audio control engineer's position. Their outputs were wired into the audio board. The table-top Ampex reel-to-reel tape recorder was settled to the right of the board, spaced comfortably between the console and the right side turntable; it was as well hooked into the console. Microphone stands were placed securely in both the main and second studios and Electo-Voice microphones too were wired into the console. "On-Air" signs and studio speakers were hung and connected. In the process of building the studios, we taught each other to interpret schematics, to cleanly solder, to use voltage/ammeters and, ultimately, to use the equipment.

Then we worked on radio signal transmission. While others worked to connect the studio signal output to the transmitters, Steve, myself and others worked to integrate the transmitters and the signal into the school's electric power. (These transmitters used the common electric current throughout the campus as radio signal carriers.) On the advice of the transmitter company techs we located the transmitters in the building elevator houses atop each campus structure. After hooking in our studio feed, we constantly took readings from the transmitters while we monitored signal reception over plugged in radios.

Unfortunately, a few transmitters were not working. As it was turning out, I was the one mostly talking with the transmitter techs trying to figure out reasons. For this and because I was the only one gleefully willing to hand deliver intractable transmitters to and from the supplier's shop in Frazer, PA, not all that far from Philadelphia(!), Steve asked me and I agreed to be the station's Chief Engineer. Finally, impedance mismatches were given the blame, fixes were made and we signed-on as WLIU 590 AM in December 1969.

We were in the dark when it came to programming, but collectively we figured that one out. We projected ourselves as progressive radio rockers. As WNEW-FM was the only Progressive Rock radio in New York City, we agreed to copy its format. Those with a finer sense of musical sound and composition heard the dynamics within and between "NEW's twelve to fifteen minute music sets." These sound and lyric connections became an aesthetic objective for our air shifts: we were creating an art form. And I made such art for three hours each of three days a week for the rest of the school year!

Our DJ patter was the only aspect of our broadcasting we developed independently of each other. Beyond artists, albums and tracks, I heard NEW DJ's broadcasting the latest news about the bands they played as well as the latest rock shows they attended or anticipated attending. The habit of reading *Billboard*, *Cashbox*, and *Radio and Record* allowed me to dish the dirt as well. However, my near poverty from being in Philadelphia so frequently precluded me from going to any shows; so, I couldn't talk about that. Yet, I didn't want to talk much, as my "thing" was the aesthetics of mixing sound and lyrics, the art form: What I had to say was "all in the mix".

Heavens, we knew we existed, but no one else on campus did. We came up with a Publicity Department. Our little group thought the best way of getting noticed was to give away a lot of stuff with "WLIU-AM 590" printed on everything. So, we purchased and distributed an array of nick-knacks including buttons and pens. But the purchase that every one of us liked too much to give away were our windbreakers: We wore them as our club jackets. Our publicity worked to draw more of our fellow students into being part of the radio station staff. Listeners were few but building as the Spring '70 semester went on.

Not a few over-nights and weekends did we spend at the radio station. Across the hall, opposite our main studio, was a large unoccupied room. We decided to make it ours and filled it with comfortable furniture liberated from other offices. During the day we slept, ate, read, and generally hung-out in our "lounge" as classes and air shifts allowed. At night, we passed the wine, listened to our station, talked music, books, poetry, politics, war, peace, civil rights, and even on occasion considered "the big questions of life."

All in all, during academic year '69–'70, the team's projection of its definition of the situation integrated us into a highly motivated cooperative task group getting it all done while having a great party doing it.

Almost every one of the team was graduating in June, leaving me with schoolmates possessing definitions of the situation very different than mine. I decided it was time to move out of school and into the industry. In the late Spring, 1970, I went to the headquarters of the American Broadcasting Company (ABC) and was placed as a Payroll Clerk in Television Production Services. It was the perfect entry level job. It lasted all of three months.

On a cold late January 1970 evening I had told my folks I was heading into radio broadcasting as a DJ. My father waited to object. He broke his silence at a family gathering a little over two months into my stay at ABC. The drink taken had reduced his inhibitions and he finally let out a fusillade of invective on my broadcasting job, my radio career intentions, and my decision to turn my back on a military academic career. I was defenseless. I could no longer sustain the work at ABC.

I went crawling back to LIU in September as a Theater major thinking that it was the closest I could get allowing me to develop as on-air talent. I re-entered WLIU but at a much diminished capacity. And I had developed a ferocious mic-fright.

As anticipated, the definitions of the situation of the school-mates now at the station were different in almost every respect from my previous team's. They projected performances of being there exclusively as members of a student club, as opposed to us who defined our participation as the proximate platform for careers. We invested ourselves in the underground art, anti-commercial, and anti-establishment mythology of Progressive Rock; they adopted the established commercial order of Top Forty radio without the tight, sophisticated humor I heard at WFIL. We approached responsibilities as the professionals we were on the verge of becoming. They could care less. We identified as lefty radicals; they were the young conservatives.

My personal definition of the situation was somewhat different than it was in the previous academic year, placing me at an even greater distance from the coalescing peer group. I had been a working professional, albeit very briefly. The performance I projected, humbly befitting the brevity of employment, was as a novice professional broadcaster returning to complete degree requirements before moving on in the industry. Further, in mid-September, 1970, I entered a relationship with the young woman who would become my life-long married partner a year and a month later; as the relationship deepened I became more intent on broadcast career development and projected a seriousness of purpose absent in everyone else.

The incompatibility of the definitions of the situation left me isolated from my college radio peers. I became a "one-man team" (Goffman, 1959, p. 85) having to perfect broadcast skills all by my lonesome, a status I was to be in from then on. I asked for a DJ shift and was given a two hour once a week show.

I spent no time hanging around the station and even less time socializing with those in charge. I arrived at least an hour before show time, pulled albums, checked the log, spun selected tracks putting the albums away after each play. I immediately left after my show and headed back to my true love. The parsimonious DJ patter during '69-'70 was an intentional device clearing additional time for the musical art form. Now, with the mic-fright, I even stumbled through letting an audience know artist, album, track, station ID and my name!

As I was outside the peer group, I no longer had the buffer between my father's demands and my own decisions. This caught me a great dilemma: On the one hand, I deeply felt I needed to comply with my father's demands to excel in academics and then graduate into a scholarly military career and on the other I deeply felt I needed to down-right disobey him excelling in broadcast skills and then obtaining sustained work in the field. The consequence of this quandary was a psychological paralysis forcing a withdrawal from classes, radio, and reality from January to August '71. I returned to classes for fall, but, withdrew for good at its end.

My wife was my new teammate. She convinced me to work through the paralysis to find a way into broadcasting. I had no access to studios, so I could not bring down the mic-fright. We decided, then, I should enter DJ work through the "back door" of the engineering department. I obtained a Federal Communication Commission First Class Radiotelephone Operator's License in November, '72. Gaining an engineering job took a while, but by the end of August 1973, I was working at WKAP, Whitehall, PA—a 5,000 watt day, 1,000 watt night Top Forty rocker. There I paid far more attention to my passions for programming, on-air performance, spot commercial copywriting, and production than to broadcast electronics, thus earning the enmity of my boss, the Chief Engineer. He fired me at the end of January 1974.

We returned to Brooklyn and I enrolled for Fall '74 at the Old Westbury campus of the New York Institute of Technology since it offered the only undergraduate radio course concentration in the metro area then. I immediately went to the closed-circuit student-run radio station, WNYT.

The DJ slots were taken, so I went to news. Seeing that I had some real world experience at WKAP, doing rip-and-read Monday morning sign-on newscasts and follow-up information gathering from listener news tips, I thought to prepare myself to become the news director on first availability by being the story assignment editor. Well, I kept on being accused of "ego-tripping" by the much younger student radio news reporters when I asked them to cover specific stories. I was definitely out of sync with my WNYT school-mates. My definition of the situation of a young radio broadcast professional returning to college to complete degree requirements before returning to the field was roundly rejected. My projection conflicted with their sense of me as just another Communication Arts undergrad like themselves. I left the radio station at the end of the semester.

Removed from any radio peers, my wife and I were left to ourselves to cope with the powerful, ever present specter of fatherly disapproval. The best way we agreed of dealing with it was to compromise with the father-induced dilemma demons: I was to accept demands for academic achievement but in MY chosen field of Communication Arts. So I did, earning honors and a Summa Cum Laude along with my Bachelor of Fine Arts degree in June, 1976.

I discovered during the two years completing the degree at NYIT that I had an innate capacity for academics. We thought it reasonable, then, that I continue with fatherly summonses and pursue an academic career in Communication/Media Studies. But, I still had a deep desire for a career in radio and a massive case of mic-fright.

Nonetheless, immediately after graduating I took a Master of Arts in Media Studies at The New School where the Program Director, Peter Haratonik, worked with me to mature my scholarship. It was at his encouragement I applied to and was accepted into New York University's Doctoral Media Ecology Class of 1982.

I went to NYU's student-run radio station shortly after completing first semester study. The definitions of the situation of the undergraduate staff and me were way out of sync. They projected performances of being the hippest of the eighties exulting in the decade's speech, music, dress, and dance. Some had occasional gigs as DJ's at CBGB's. I, on the other hand, was projecting the performance of an "aging Sixties' hippie" keeping his hand in college radio while working on his professorial degree. Besides, I was already an adjunct media college instructor. I continued being a team of one at WNYU.

My first thought for what I wanted to do there was the journalism initiated at WNYT. So, I petitioned the then news director to be accepted in his department as a newscaster. Passing on my fitness, the undergraduate tore into my "old fashioned" news copywriting sternly saying that WNYU was not National Public Radio and I needed to take no more than forty-five seconds to get a story told. My teaching afforded me a good grasp of radio news copywriting enabling me to accommodate the young man's demands. Satisfied on having met his standards, he assigned me newscasts on the closed-circuit AM, used to train, polish, and prepare for ascendency to the over-the-air FM. It turned out to be a glorified rip-and-read from local newspapers. Still, it put me in front of a microphone letting me work through the mic-fright. On my own, I struggled with the dilemma demons, hemming and hawing through each five minute cast. The demons won. I removed myself from the news department. And then I met Hugh Foley, WNYU's program director.

Hugh was older than the others at the radio station, but not as elderly as I. He, too, had some professional experience. The proximity in age and experience created a foundation for friendship and a collaboration in a two-man team. We projected three professional goals: to keep moving forward in our radio careers; to excel in programming and production; and to erase any impediments to a relaxed on-air performance.

Hugh's enthusiasm for our project kept me on the AM as a DJ as long as I wanted to be, as long as I needed to be, and allowed me the confidence to devise a plan to overcome the immediate challenge of the open mic paralysis. It centered on applying to myself strategies I had developed to instruct radio broadcast undergraduates. There were two stages in this approach: Stage 1-working on voice, copy reading and on-air performance; Stage 2: forgetting all done during Stage 1, relaxing and having fun selecting, listening to and talking about music and about any other topic which interested me.

Freeing the Natural Voice by Kristin Linklater (1976) became my bible in voice development. Linklater's vocal exercises, especially the vocal warm-ups prior to every DJ shift, proved themselves when we heard my deepening, resonating, and relaxed voice on tape-recorded air-checks. Listening to the tape-recorded, off-air copy reading under her performance advice continued developing a more relaxed, error free reading. The more I read, the better I became, especially at sight-reading. The needed on-air progress was happening but slowly, principally due to *ad lib* stumbling. Extemporaneous open mic patter showed the hold the mic-fright still had. Yet, even here I was improving.

As I was moving to Stage Two relaxation and enjoyment, Hugh gave me his weekly over-the-air FM show, *Back Roads*. As he put it, *Back Roads* should be the fastest moving sixty minutes on New York City radio. Well, I thought, this is the stepping stone back into the industry and this time as the DJ I had always wanted to be.

However, all the self-help and support from Hugh left untouched the mic-fright's psychological foundation. Hiding within the hobby of AM development radio, while doing the academics my father commanded, satisfied the dilemma demons enough for me to work through them, advancing my radio skills. But being over-the-air in the largest major radio market roused them. With every *Back Roads* show I had to fight them telling me to fail, fold, and run away. The old psychological paralysis reasserted itself. In well under a year on the FM, I removed myself from *Back Roads*, WNYU and New York University leaving the Doctorate incomplete at the All But Dissertation point.

REFLECTIONS

Fatherly demands had me as an academic in Military Science; I became an academic in Media Ecology. Fatherly expectation of my being an academic was to teach; I became a part-time Communication/Media college teacher with a passion for instruction. I never was a professional radio broadcaster after WKAP, never was a professional radio DJ at all.

Inclusion within a college radio peer group provided a barrier between this individual's broadcast career decisions and parental career demands for other vocational aims. But, being by my lonesome, even with a thoroughly committed, loving and caring, life-long partner, was insufficient to deflect the effects of parent over-involvement.

There may be some who can by themselves overcome parental power, but this study suggests otherwise.

REFERENCES

Bergen, R. J. (2006). *Family influences on young adult career development and aspirations.* Retrieved from http://digital.library.unt.edu/ark:/67531/metadc5480/m2/1/high_res_d/dissertation.pdf

Clutter, C. (2010). *The effects of parental influence on their children's career choices.* Manhattan, KA: Kansas State University. Retrieved from https://krex.kstate.edu/dspace/bitstream/handle/2097/3951/ChanceClutter2010.pdf?

Creamer, E. G., & Laughlin, A. (2005). Self-authorship and women's career decision making. *Journal of College Student Development, 46*(1), 13–27.

Ferry, N. M. (2006). Factors influencing career choices of adolescents and young adults in rural Pennsylvania. *Journal of Extension, 44*(3), 1–6.

Goffman, E. (1959). *The presentation of self in everyday life.* Garden City, NY: Doubleday Anchor Books.

Linklater, K. (1976) *Freeing the natural voice.* New York, NY: Drama Book Publishers.

Mitchell, L. K., & Krumboltz, J. D. (1990). Social learning approach to career decision-making: Krumboltz's theory. In: D. Brown & L. Brooks (Eds.), *Career choice and development: Applying contemporary theories to practice* (pp. 145–196). San Francisco, CA: Jossey-Bass.

CHAPTER SIX

Lessons Learned

How College Newsrooms Prepare Students for the Professional Newsroom and Beyond

LEIGH LANDINI WRIGHT

Long before the Knight Foundation's Eric Newton advocated a teaching hospital model for journalism education, college students toiled away in campus newsrooms to produce a daily or weekly newspaper. Just like professionals at weekly or daily newspapers, the hours were long, the pay low (or in some cases, nonexistent) but the satisfaction of breaking a big story, high.

Their stories didn't come from a textbook, but rather the news and events happening on their campus or in the community. News about university budget cuts, news about athletes running into legal trouble, news about crime happening on campus. The list could go on. The students worked as legitimate journalists in a university setting to get their stories every day or every week. They asked the same questions as the professional journalists, often with minimal supervision from an adviser, and they experimented with story structures and writing styles, just as the professional journalists did. The only difference came with their hours, sandwiched around classes, or a lower paycheck, or for some students no paycheck as their newsrooms ran as class practicums.

First, let's examine the teaching hospital model of journalism education that Newton advocates. In an address to Dutch educators in November 2013, Newton, then the senior adviser to the president of the Knight Foundation, outlined his thoughts about the teaching hospital model for journalism education. He defined the concept as an "ancient idea that people learn by doing," a quote that he attributed to the Confucian philosopher Xunxi who wrote down a Chinese proverb: "I hear and I forget; I see and I remember; I do and I understand" (Newton, 2013a).

In his *Searchlight and Sunglasses* e-book, Newton explained the teaching hospital model as "A model of learning-by-doing that includes college students, professors and professionals working together under one 'digital roof' for the benefit of a community" (Newton, 2013b).

Although Newton acknowledged that laboratory news programs at the University of Missouri provide students with on-the-job training, he envisions a teaching hospital model as one that combines the notions of students doing journalism alongside professionals who mentor them to improve their product. Additionally, these elements are combined with professors who deliver knowledge and raise questions, academics who conduct research projects, innovators who devise new tools, and with everyone working together to inform and engage the community (Newton, 2013b).

The debate between how intense of a role professional skills should play in journalism training is one that has existed since the beginning of journalism education. Cornell University and Washington College (now Washington and Lee University) both offered some of the earliest classes in journalism in the 1860s; Robert E. Lee, the famed Confederate general who became the president of Washington College, helped establish 50 scholarships for young men who wanted to work in journalism (Folkerts, 2014). However, as the 1800s continued, newspaper editors became divided over whether they wanted journalists trained in their shops or those trained in the academy. *New York Tribune* editor Whitelaw Reid said two-thirds of the reporters he knew in the New York media were college-educated, but later Horace Greeley, founder of the *Tribune*, declined to hire college graduates (Folkerts, 2014).

As journalists and educators argued about the role of professionalization in the curriculum, schools such as the University of Missouri, began offering a journalism program heavy on skills, ethics and history (Folkerts, 2014). For example, The University of Missouri School of Journalism opened on September 14, 1908 and immediately published the first issue of the *University Missourian*; 106 years later, the newspaper is now known as the *Columbia Missourian* and is edited by professionals yet staffed by students (Francisco, Lenhoff, & Schudson, 2012). The publication does not pay students because the staff is drawn from eight journalism skills classes and students receive academic credit (Francisco et al., 2012). Missouri's model sought to combine the lab and lecture with the lab supporting the classroom lessons (Folkerts, 2014). Other schools combine this type of model of learning by doing with their classes, but the student newsrooms may be independent of the classes and students are paid for their hours.

The Missouri Method isn't far from what Newton advocates as the teaching hospital model and it's one that many campuses have copied with student-run publications or news stations. But his model adds in the layer of innovation and research that may not be present in all college media programs.

Newton as well as Missouri's journalism school founder Walter Williams have advocated the model of learning by doing. Whether schools adhere to the Missouri Method or Newton's teaching hospital model, one thing remains the same. Journalism and media are fields in which people learn best by doing the actual job of a journalist rather than working out of a textbook or workbook full of archaic exercises. Mistakes are made. Corrections run. Lessons learned. Students learn by doing real journalism for a real audience.

COLLEGE NEWSROOMS PREPARE STUDENTS FOR PROFESSIONAL ROLES

Elizabeth Johnson, a reporter at the *Sarasota (Florida) Herald-Tribune*, said she enjoyed putting the skills she learned in the classrooms at Murray State University into practice first as a sports reporter and sports editor and later as editor-in-chief of *The Murray State News*. Johnson graduated in 2012.

> The great thing about working for college media is that you actually get to do what you're learning in class. But not everything is textbook, so you have to take what you learn and adapt it to whatever you're reporting. Maybe you need to be a hard-nosed watch-dog, maybe you need to be friendly to gain trust. Maybe you need to write down everything the person says, or maybe you need to use a recorder so you can maintain eye contact. Maybe you need to stick to your prepared questions to keep a source from veering from the topic at hand, or maybe you need to go off-script. (E. Johnson, personal communication, December 29, 2014)

Johnson said those lessons help her now as a cops/breaking news reporter. She live-tweets from crime scenes, news conferences, and court proceedings; she writes quick updates for the web before she delves into an in-depth story for the print product.

Jennifer Smith, now a sports writer at the *Lexington (Kentucky) Herald-Leader*, agrees. Smith worked in a variety of roles, including editor-in-chief at the *Kentucky Kernel*, the University of Kentucky's campus paper in the 1990s but left college early for a career. Smith returned to school and graduated in 2008.

> You can rewrite press releases into news stories and conduct pretend interviews in the classroom setting a million times, but still not be fully prepared for going out and asking someone a question that he/she doesn't want to answer, grabbing up pertinent facts and then writing a story on deadline that is understandable for the reader. At the college paper, you're learning to do all of those things in a real-world scenario with real-world implications. (J. Smith, personal correspondence, January 1, 2015)

For instance, Smith said if a student incorrectly writes a "fact" in an in-class story, the professor lowers the grade, but if that incorrect fact is published in the campus paper, readers will contact the editor and reporters to complain. Editors will demand fact-checking and demand deadlines, she said. "There is nothing in the classroom setting that can prepare you for the idea that what you write can make a difference or for the backlash that you might receive over what you've written" (J. Smith, personal communication, January 1, 2015).

Hillary Copsey, who graduated from Ohio University in 2003, said most of her education came at the campus newspaper, *The Post*. She learned skills and knowledge such as grammar and law in the classes, but it was the constant practice that helped propel her career from being a reporter at *The Winchester (Virginia) Star* newspaper to the Scripps Treasure Coast Newspapers in Stuart, Florida, where she served in a variety of roles including features editor and social media editor for 10 years. While in college, Copsey worked on several reporting beats at the five-day-a-week student paper and was an editor for the weekly summer product. She also worked as a copy editor at *The Post*.

> I was always a self-starter, thanks to my time at the student-run paper. That and the reporting skills I learned on the job enabled me to find stories on my own and quickly write and edit my own copy, vital skills in the enterprise reporting positions I wanted and then moved into. (H. Copsey, personal correspondence, December 23, 2014)

For Jacque E. Day, the lessons learned at *The Penn*, at the Indiana University at Pennsylvania, became her journalism education. She was an English major who joined *The Penn's* staff as a feature writer and worked her way into the news editor's job her senior year in 1993–1994.

> By the time I hit the news desk, I had some gut instinct for what went into a news story, the level of research required, how to balance sources and quote in context. I still had a lot to learn, but I was able to train the team and get them a little further along in their own skills. (J. Day, personal correspondence, December 31, 2014)

Gill Welsch, who started working as circulation manager for *The Student Printz* at the University of Southern Mississippi in 1983, said his only regret from working in student media was that he didn't start sooner and develop a better clip file. He rose through the ranks to copy editor, news editor, and columnist before graduating and taking a job with *The Boliver Commercial* in Cleveland, Mississippi. Welsch now teaches advertising at Murray State University.

> Working in the newsroom taught me how to manage myself so that my work would be submitted by deadline. This was particularly challenging for the Thursday paper. We had to start working on it Tuesday morning. We basically had a day to get it finished because the paper went to bed Wednesday afternoon. That schedule proved to be helpful when I started working at a five-day-a-week paper after graduation. We were always on deadline. The college paper's publication schedule helped me prepare for that. (G. Welsch, personal communication, December 30, 2014)

COLLEGE MEDIA PREPARES STUDENTS FOR TEAMWORK AND LEADERSHIP

The college newsroom also provides lessons in teamwork and collaboration, skills that reporters and editors need for the professional setting. Stephanie Elder, a 2004 graduate of Murray State, worked as a sports reporter for the *Murray State News* and a producer for MSU TV-11 before going on to broadcast production jobs.

> Once I took my first role as a news producer, I used those skills of writing, reporting and teamwork to continue to develop as a journalist. After several years in the business, I decided to move into the public relations sector. Having worked in the media gave me a competitive edge in the field of PR. I continued to develop my journalism skills of writing and producing but in a different capacity. I was now writing news releases and creating videos and website content. Student media allowed me the basic skills needed to continue in the many roles that I have taken in the field of journalism. (S. Elder, personal communication, December 29, 2014)

Johnson learned that people have to let go of their perfectionism, and she had to trust people to do their jobs as she had been trusted by her adviser to do hers as editor. The goal becomes to produce a top-quality product that's accurate and informative. As a reporter, she learned that she had to turn in clean copy, but as a college editor, she had to fact-check her staff's copy as well as provide them with ideas and inspiration to do their job, not do their job for them.

> Reporters tend to be selfish people. There is a big story, and we want only our names on it. Sometimes it works out that way. Sometimes we share the byline or the awards. You have to learn that there is nothing wrong with that. Some of my best stories have double bylines, and I'm OK with that because working with another reporter can be a challenging and inspiring experience. (E. Johnson, personal communication, December 29, 2014)

The college newsroom also provides an opportunity for students to develop leadership skills. For Johnson, she learned that editors have to communicate, whether they're juggling college classes or professional assignments, and they have to work smarter in all cases. She explained, "You have to learn how to be a friend and a leader. In college, it's extremely uncommon that managers are in charge of classmates or friends. You have to find that professionalism" (E. Johnson, personal correspondence, December 29, 2014).

Day, the lone English major at *The Penn*, recalled that she had to become creative to manage a staff of paid and volunteer writers for the news desk. The students easily could have made more money in other jobs, so she worked to motivate them to find relevant stories for the readership and trained the younger students to emerge as staff reporters.

For Smith, the on-the-job training at the *Kernel* helped her to learn how to manage her peers while also sitting beside them in classes. The criticism that she received as a student editor helped her learn how to take criticism as a professional.

> It helped me to understand that you build consensus when possible, but make solo decisions when they're warranted. Being editor-in-chief taught me how to manage different personalities (think sports editor vs. arts columnist) effectively. I had to learn how to work within budget constraints (how to pay everyone and the column-inch types of budgets) and under deadline constraints. (J. Smith, personal correspondence, January 1, 2015)

But even with those challenges, Smith said she developed friendships that have endured the frustrations of the college newsroom. They celebrated the big and small victories. Major stories. The paper simply getting out every day.

> My staff even had to contend with getting a national championship section out as well as a special snow edition when a blizzard closed campus. Those are things that you will never, ever experience sitting at a computer in a journalism classroom. (J. Smith, personal correspondence, January 1, 2015)

Campus newsrooms change staff frequently as students graduate or change roles within the newsroom. The constant change within the *Post*'s editing staff at Ohio University helped Copsey become more adaptable while still retaining her own voice in her writing. "I learned to work collaboratively with a variety of people. I learned how to do as much as possible with the few resources I have" (H. Copsey, personal correspondence, December 23, 2014).

Whether the campus paper is published daily or weekly, students gain the confidence and ability to learn leadership skills. Smith said she learned that she had to set reasonable goals for her staff and that she couldn't expect them to do something that she didn't want to do.

> I learned how to be seen as a leader at the campus level: meeting with administrators while writing the administration beat, becoming a voice of students via our staff editorials, dealing with the grief of a student being murdered and covering the funeral, then dealing with the aftermath of fraternity members coming into your office for covering the funeral. College journalism prepared me for so many situations where I have to be a strong leader, including standing up for myself or my work when people disagree. For instance, rising above the fray in social media situations, acting like a grownup and a professional even when others are not. (J. Smith, personal correspondence, January 1, 2015)

Johnson said editors must communicate, which promotes leadership and teamwork, and the communication keeps the workers on track. "Inside the newsroom, there must be a mutual respect among all employees. You have to smartly choose which battles are worth the fight and where compromise is the best option" (E. Johnson, personal correspondence, December 29, 2014).

Elder's campus experience as a news producer for the TV station gave her the ability and confidence to step into a leadership role soon after graduation for production jobs. "I found this to be tremendously influential as I decided to go into producing/newsroom management" (S. Elder, personal communication, December 30, 2014). Elder has since left the media business to focus on pursuing an Ed.D., as well as special education certification for the public school system, but she teaches advertising, public relations, and broadcast reporting classes part time.

PROFESSIONALS DEMAND DIGITAL SKILLS IN JOURNALISM CURRICULUMS

While most professionals seem pleased with their campus media experience, all expressed a desire for current journalism education to do a better job of integrating social media, multimedia, digital storytelling, and informational graphics into classes and newsrooms as those skills are needed in the professional world. Copsey participated in an AEJMC/Scripps Howard program with the author of this book chapter to bring professional skills in social media to the author's classrooms and campus newsroom. The author participated in a two-week visiting professorship in which she observed and participated in reporting using social media and using social media engagement strategies. Those observations and experiences translated into real-world social media writing exercises in the classroom.

Elsewhere, schools such as UC Berkley, New York University, and the University of Oregon are launching digital news practicums to help students develop the cutting-edge skills they will need for tomorrow's newsrooms (Madison, 2014). These practicums range from hyper-local news sites to a news laboratory. They provide students with the ability to get their stories out to a larger audience, but still, students are learning by doing and doing this with the cutting-edge skills that news organizations increasingly are demanding.

BEYOND THE NEWSROOM

Journalists who work with student media whether as a consultant or as an instructor should rely on their background and professionalism to stress the importance of creating a top-notch product. For example, Day took her experience from a Chicago weekly newspaper and a National Public Radio affiliate in Kentucky with her when she transitioned into teaching at Northern Kentucky University, where she served as adviser for a year to the campus paper and the campus radio station.

> Drawing on my background in news radio, I guided the two organizations to a more cooperative relationship, and by the time I left, the station had better technology, and they'd created a news director position to integrate news into the programming. It was a lot of fun for me to coach the journalism students in the art of producing radio news. (Day, personal communication, December 31, 2014)

As several of the journalists mentioned, the experience gained in the campus newsroom far exceeded what they learned in classes. Day merged her journalism practicum classes with the campus paper and asked the assignment editor to give assignments to the class. She explained, "Then in the class, we'd cover practical aspects of reporting that would hopefully help them envision their stories—for instance, what sources to seek out, how many and how to balance the voices" (J. Day, personal communication, December 31, 2014).

She learned that the same frantic energy of the newsroom didn't translate as well into the copy editing and page design classes. She said the students were not accustomed to the pace of a newsroom at that point and she had to slow down to allow the students to better process the information. In the end, she integrated the skills of editing and fact checking with headline writing to build a full newspaper layout. "They got the whole experience across a semester, whereas in the newsroom, the responsibilities are divided up and the product comes together very quickly" (J. Day, personal correspondence, December 31, 2014).

LEARNING NEVER STOPS

Even after college, journalists and media professionals agree that the learning didn't stop. Johnson reflected on several stories she wrote a year ago and says now she would have handled them differently with new skills and knowledge gained in the field. She's also honing her interviewing skills and experimenting more with narrative prose.

> What's important is that I'm not afraid to try. In college, I had struggles. I got burned a couple of times. People felt burned by me a couple of times. I dealt with hostile sources—including the university president. I had [to make] corrections (E. Johnson, personal communication, December 29, 2014).

Johnson has experimented more with her writing style and the voice in her pieces, particularly analyses. Day eventually transitioned from news reporting to literary journalism, but she kept her writing firmly rooted in fact.

> Now I'm on the staffs of two very different publications—a collegiate alumni magazine and a literary journal. I apply the same rigid standard for fact-checking to those as I would any hard news story. And I don't ever feel confined to stiff or dry writing. (J. Day, personal communication, December 31, 2014)

Welsch acknowledged that many students who start out in journalism transition to other fields throughout their careers whether it's public relations, law, or fields outside the media. But with the training as a journalist to ask questions and think critically and analytically, they can translate their skills to other jobs. "When you know how to learn and when you know how to think on your feet, you can do pretty much anything" (G. Welsch, personal communication, December 30, 2014).

Whether it's learning new writing styles or learning new digital and social media tools, media professionals cannot stop the learning that happened in their college newsrooms. Those lessons learned translate into a lifetime of learning and doing journalism.

REFERENCES

Folkerts, J. (2014). History of journalism education. *Journalism & Communication Monographs*, *16*(4), 227–299.

Francisco, T., Lenhoff, A., & Schudson, M. (2012). The classroom as newsroom: Leveraging university resources for public affairs reporting. *International Journal of Communication*, *6*, 21.

Madison, E. (2014). Training digital age journalists blurring the distinction between students and professionals. *Journalism & Mass Communication Educator*, 69(3), 314–324. DOI: 1077695814532926.

Newton, E. (2013a, November 22). The "teaching hospital"—a goal for journalism education. Retrieved from http://www.knightfoundation.org/press-room/speech/teaching-hospital-goal-journalism-education

Newton, E. (2013b). *Searchlights and sunglasses: Field notes from the digital age of journalism.* Retrieved from http://searchlightsandsunglasses.org

SECTION THREE

Case Studies

CHAPTER SEVEN

Student Newsmakers

A Case Study in Socialization and Professionalization

KIMBERLY MELTZER

Many professional journalists were at one time members of student newspaper organizations on college campuses. This was where they began honing their skills and learning the roles and practices of journalists. While many studies of professional journalists and news organizations have been conducted, there are few focused on non-professionals, or student journalists, and of this body of work, even fewer delve specifically into the routines of, and relationships between, members of student newspaper staffs.

This project's charge was to understand what the social and professional atmospheres of college student newspaper organizations are like. In particular, this study is concerned with where student newspaper staff members get their ideas about what the role of a journalist should be and which practices to follow; do they emulate models found in the professional journalism world? It also asks why these students devote so much time and effort to this task, what draws them to the activity in the first place, and how they negotiate the tensions between their jobs at the paper and their responsibilities to schoolwork and social life. Finally, this work looks at how new members are instructed about the norms and rules of the paper.

STUDIES OF NEWS ORGANIZATIONS AND JOURNALISTS

In the realm of journalism and newsroom scholarship, there are a few studies that have been central to the field and have impacted research through the present. In the tradition of research on the sociology of news, this paper follows in a similar vein to that of Tuchman (1978), Gans (1980), Fishman (1980), Breed (1955) and more recently Klinenberg (2003), Boczkowski (2010), Usher (2012a, 2012b), Bivens (2014) and others, albeit on a much smaller scale, as the length of my time in the field was much briefer and the scope of the research much narrower. The intent was to conduct an ethnographic study of a news organization, but unlike these other scholarly works on professional news organizations, this one would be of a college newspaper run and staffed by students.

Although it is a smaller body of work, there have been a number of studies conducted on student journalists and student news organizations. Marston (1999) advocated for studying college news-workers, and conducted an ethnographic study in 1996 of student journalists at the *Daily Iowan*. She found that:

> [C]ollege news-workers love[d] their work and got a great sense of purpose from it but [were] already harming their bodies through lack of sleep, lack of food and the development of painful, stress-related illnesses due to overwork. This … finding may not be too surprising for a college student population, but these news-workers are at a university—an educational environment that is supposed to be preparing them for their careers. (p. 267)

Marston's study looked at college news-workers through the lens of feminist and cultural theories to understand how these "college news-workers [were] beginning to be socialized into the American journalistic work culture in terms of viewing (or not viewing) their bodies in relation to their work." Her ultimate goal was to explain why journalists continue signing up for an occupation that could permanently disable them through repetitive strain injury that the author herself suffered from.

Other studies of student journalists have yielded both positive and negative characteristics. Bodle (1997) compared the newswriting in six mid-western community daily newspapers to the writing in six daily student newspapers and found that in the areas of readability, interest level and thoroughness, the news in community dailies was not significantly better than that produced by students. Watts and Wernsman (1997) found that school administrators were generally dissatisfied with student reporters' preparation and professionalism in producing stories about the administration. McDevitt, Gassaway, and Perez's study of civic journalism (2002) found that in the transition from classroom to newsroom, newfound autonomy and cultivation of a professional persona precluded attempts at community engagement by journalists.

None of these studies, however, takes as its aim the understanding of the processes through which student journalists are socialized into the newspaper organization, and how their status as students affects their roles as staff members. This study seeks to do chiefly that.

METHODOLOGY

This study employed an interpretive qualitative approach (Gunter, 2000) to analyze the discussion of student journalists at one college newspaper. I was granted access to the campus student-run newspaper and was allowed to spend time at the paper's headquarters, attending and observing meetings and informally speaking to staff members. I clearly identified the nature of my intentions to observe the meetings and events as an academic researcher.

The majority of staffers with whom I spoke at length were editors. In total, I had conversations with 10 of the paper's 23 editors, in addition to several other staffers. According to the editorial page editor, there were approximately 10 columnists, 20 beat reporters, 20 general assignment reporters, 5 copy assistants, 15 sports writers, 20 photographers, and 20 arts and entertainment magazine writers. He was close in his estimations to the number of staffers listed in the April 28, 2004 issue of the paper as "Spring 2004 staff." The exceptions were that the paper listed twice as many sports reporters and over three times as many copy assistants. Additionally, the paper listed 33 photographers. Relatively consistent with Wearden, Hipsman, and Greenman's findings about racial diversity in the college newsroom (1990), the *Daily Collegian* has few minority staffers and even fewer minority editors. Of seventeen editors who were present at the editors' meetings, there was one African American female, one Asian female, and the rest of those present were white. This could be reflective of the undergraduate student body at the University. As the senior sports editor said, "we don't turn anyone away who wants to write for us, so it's a self-selecting group" that chooses to work for the paper.

I made visits to the paper, which I will refer to by a pseudonym, the *Daily Collegian*, on several occasions, and also accompanied a reporter as she covered a live event. Additionally, I attended a forum open to the public that was held by the paper's executive editor. My fieldwork was conducted over a two-month period in the spring of 2004. Since this was before the height of the revolution of online and mobile journalism, it offers a snapshot of what the experience of college newspaper journalists was like at the time. I chose to focus exclusively on the editorial side of the newspaper, rather than the business side, but did find that the two work largely in isolation from one another.

Although I had a tape recorder with me on several occasions, I found that due to the nature of my interactions, which were mostly informal chats in the hallways and offices of the paper, with staffers coming and going, it was not feasible to use the device to tape conversations. As a result, the sessions were not tape-recorded. Although I probably would have been given permission to do so, I believe that taping may also have distracted the meeting participants and affected their behavior. Instead, careful notes were taken from the start to end of each interaction. Reports of conversations have been paraphrased from the exact quotations that were made at the time, but effort has been taken to be as accurate as possible and ensure that the tone and content of the messages very closely reflect what was said. In order to protect the identity of the speakers, substitute names or only titles are used. *The Weekly Collegian,* a version of the paper which gets sent to alumni and parents who subscribe to it, is published out of the same offices and put together by one person—the editor-in-chief. There are no ads in the *Weekly Collegian*—it is supported by subscriber revenue only. The editor-in-chief selects stories to include from the *Daily Collegian* each week, and runs this by the *DC* managing editor. The *WC* gets sent out to its subscribers on Tuesdays.

ANALYSIS

From observations and discussions with members of the newspaper staff, several organizational routines and characteristics surfaced. I clustered these according to several themes. These are (1) why students first decide, and continue, to work for the newspaper; (2) how their work at the newspaper gets done, which includes balancing of schoolwork with newspaper staff duties, and formal norms and routines of the paper; (3) the ways in which they cultivate and work within professional journalism culture, through comparisons to other college newspapers, professional newspapers or journalism societies; and (4) the informal relationships that exist between student journalists within the organization. I will describe each of these themes in detail and provide examples.

Why Students Work for the Newspaper

When asked why they were willing to spend many hours working for the paper on weekends, late at night, and when other pressures bore down on them, members of the student newspaper responded to my question, and sometimes told me unsolicited, that the perks that came with the job, the preparation that working for the newspaper gave them for their careers, their natural interests in current events, the respect earned from their peers, and the camaraderie among the paper's members, were all reasons.

The Sports section editors said that their willingness to devote all this time and energy to the paper was at least in part accounted for by the press passes they were granted for different types of games, the free travel for which the newspaper paid in order for them to cover games, and the respect they received from classmates. The senior sports editor said that "everyone [the other editor chimed in that 97% of the student body] reads the paper, and there is respect that goes with working for the paper" [then the third editor said, "although most people take it for granted"].

On the press pass front, I commented to the reporter I accompanied to cover a live event, that it was nice to be able to bypass the entire huge line of students waiting to get in to the event on that rainy night, and go right up to the door where she was waiting with a press pass for me. She led me to the press section all the way in the front of the auditorium on the right side. There were reserved seats for us, and we joined two *DC* photographers and a reporter from the University Almanac, as well as another man who though unidentified, appeared to be a reporter taking notes. She told me the event was sold out, but we had front row seats. The sports editors told me that "it's cool that when we go to cover big games, we get to sit in the press box next to reporters from Comcast Sports Net, UPN and the *New York Times*."

I also learned that editors are paid small stipends for their work at the paper, but that this only amounts to a few dollars per hour—"just enough to get coffee while working or to make up for missed meals" due to being at the paper, according to the *WC* editor-in-chief.

The executive editor of the *DC* was a business major, interested in advertising. He said that many of the students who work for the *DC* actually do not want to be journalists. One of those students was the senior sports editor, who offered an alternate reason to career for his work at the paper: "I wish this was my full-time job, but I wouldn't want to do it as a career," but "of course it looks good on your resume." He and another sports editor liked the business aspects of sports better. The senior sports editor said he would not want to be a sports reporter who covers a particular team. Consistent with this, he started a column in the paper about issues in sports, such as changes in rules. However, some of the student journalists were interested in careers in journalism and thought of the paper as a stepping stone. The third sports editor who was a senior said, "I wouldn't spend all of my time here if I didn't want to do it for a career." Although it is unusual for a senior to be an editor, since this student had transferred from another college and wanted to be a sportswriter when he graduated, he still got the job. As he prepared to graduate and look for jobs, he sent his *Daily Collegian* clips to prospective employers.

When I asked the *Weekly Collegian* editor-in-chief why people work so hard at the paper, she said she thinks they do it "because it's fun—we make jokes in the story meetings and bond at night while we're working." She also liked knowing what was going on. She said she didn't watch any other news, but she knew the news from working at the paper. She also found things out before everyone else. The example she provided was that when the new University president was nominated by the board of trustees, the newspaper's staff knew hours before anyone else. She told her roommate the night before the story ran in the paper that something big was going to happen but she "couldn't say what."

For several *DC* staff members, working for the college newspaper was a continuation of an interest generated in high school. One reporter told me she had been the editor of her high school paper, while the *WC* editor-in-chief had been design editor of her high school paper and wrote for it as well.

How Student Journalists Work: Balancing of Schoolwork with Newspaper Staff Duties

Unlike professional news organizations where being a journalist is usually the sole full-time job one performs, in student news organizations, student journalists must juggle their duties at the newspaper with many other obligations, such as schoolwork and other extra-curricular activities. This juggling act was evident through many of the interactions I had with staff members of the paper.

On two different occasions when I emailed the executive editor to ask permission to visit the paper, he wrote back okaying my visit, but saying that he wouldn't be there because he'd be at a study group for a final exam or actually taking an exam. The editor-in-chief of the *Weekly Collegian* told me that she was a bioengineering major and in the school band. As a freshman, knowing she wouldn't have that much time, she started as a design assistant. Then, at election time, she ran for design editor, but made it known in her letter of intent and interviews with the board that she didn't have that much time, so she was happy to become the *Weekly Collegian* editor because it only required her to work at the paper Thursday and Monday nights, and sometimes over the weekend. Similarly, the reporter I accompanied to cover the live event told me that when the beats were being assigned, she told them she wanted a slower one that didn't require writing all the time because she was very busy with other things, so they gave her the business school beat. During one of my visits to the paper, the senior sports editor was writing a story for the next day's paper that another reporter was supposed to write, but had to cancel because he had too much to do for final exams and papers. The senior sports editor said this doesn't happen often.

How Student Journalists Work: Formal Norms and Routines of the Paper

Although it is not a professional news organization, the student newspaper has established institutional procedures, policies, and schedules to ensure that production is conducted in an orderly fashion and that they meet their deadlines. The formal norms and routines are built into systems of acquiring new members at the beginning of each academic year, progressing through a hierarchy of reporter status, maintaining daily and weekly schedules of meetings and deadlines, conducting elections for editors every fall, and codified relationships between departments.

Acquisition of New Members Through Interest Meeting at Beginning of Each Year; Progression from General Assignment Reporter to Beat Reporter to Editor

The reporter I accompanied to cover a live event told me that she knew she wanted to write for the paper right when she started at school, and at the beginning of the year she went to a meeting to become a general assignment reporter—"that's how you start." She wrote 5–6 stories during the semester. Then she became a beat reporter. Beat reporters are expected to write three stories per week on average. The *Weekly Collegian* Editor-in-Chief said she also first went to an interest meeting at the beginning of her freshman year. The senior sports editor and copy editor joined the paper together their freshman year. They went to the interest meeting that was advertised on campus. Beyond being promoted from general assignment reporter to beat reporter, there is also a hierarchy of events within beats. For example, in sports, according to the senior sports editor, "men's basketball is at the top—the best—there are lots of people on that beat. Crew is at the bottom."

According to the executive editor, training of beat reporters takes place at the beginning of each semester. Generally, there are no regular or official training sessions. Training occurs mostly by having new reporters shadow veteran staff members.

Elections and Election Protocol—How One Gets to Be An Editor

The newspaper operates on a January to December editorial calendar year. Elections are held in November, and replacement happens around Thanksgiving. The editorial page editor told me that the summer is the time when reporters have a chance to prove themselves. Last summer, he was able to try out being editorial page editor, learning how to do layouts on the computer. Then when it came time for elections, everyone who wanted the job had to meet with the editorial board and explain how they would do a better job than the present editor, or tell what was good and bad about the way the old editor did things. When the new editors are elected, they all meet together for a sort of initiation with champagne.

Daily and Weekly Schedule

On my first visit, I found out that the *DC*'s production week is Sunday through Thursday for the most part, since the paper is only put out for weekdays. The business staff works during the day until 5 p.m. The executive editor said, "[the business staffers] work during the day—they're out by 5 p.m. There's very little interaction [with the editorial side]." The editorial staff doesn't get in until around that time. They stay until 2 or 3 a.m. The editorial page editor said that the executive editor is at the paper all night sometimes, but he made the executive editor leave at midnight the previous night and that was "a big accomplishment." On average, the editorial page editor said he works about 35–40 hours per week at the paper.

When I asked the *Weekly Collegian* editor-in-chief about staying late at the paper, she said that the pages usually don't go out until 2 or 3 a.m. She thinks the technology actually makes it take longer, "'cause they figure if they have the ability to make it better, it's worth it to stay a few more hours." She said they probably got out earlier before there were computers, but "you should see" the poor quality of photos in those days. "Now the photo department has color correction … they'd never allow the poor quality stuff from the pre-technology era now."

Relationships Between Different Departments

At the live event I accompanied one reporter to, it was apparent that the photographers and the reporter did not work together. I ask the reporter I accompanied about this and she confirmed that they go to the event and do their work independently from one another. However, it is the photographer who writes the captions. This separation between text and photographs was further evidenced by the sign posted in the *DC* office which read, "Photo Request by 8 p.m. Photo requests for Fri, Sat, Sun are all due by Thursday at 8 p.m." The different section editors place orders with the photo department according to their projected needs for the week. When the photo department cannot fill a request, alternative arrangements are made. In one instance, the photo editor told the senior sports editor that she did not think she would have a photographer available to cover an upcoming event at a competing university the next day since it was final exam time and the weather forecast called for rain so the game would probably be canceled. The sports editor asked if she could call the other school's paper to find out if they would have someone taking pictures from whom the *DC* could obtain one. The photo editor obliged.

Similarly, the design department works on all of the sections of the paper, even though the sections also lay out their own pages. One of the sports editors used to be the design editor, so he lays out the sports section.

The executive editor was obviously conscious of the firewall that is supposed to exist between the editorial and business sides of the paper. This was apparent when it came time at the end of a nightly editors' meeting to ask the group who the *DC* should contract with to provide them with food each night while they worked at the paper. The executive editor said, "We don't want someone for next semester who advertises in the paper." Incidentally, the group decided to go with a pizza company.

Nightly Meetings (Editors' Story Meetings; Executive Board Meetings)

Every evening at 6 p.m., an executive board meeting takes place with the executive editor, managing editor, and heads of the business side of the paper. Every evening at 6:30 p.m., there is an editors' meeting where the editors critique that day's paper and discuss stories for the next day. Following is an account of one of these editors' meetings:

As the meeting started, led by the executive editor, all of the editors flipped through the pages in unison. They began with the front page, then moved to page two. Comments were made from around the room:

> "I liked the cartoon." "I don't think the story on the new design school website merited a story," said the internet editor. "For the map on the front page, I gave [the writer responsible for it] six locations and she got one of them wrong," said the campus news editor. "We did get blasted for publishing letters from [the university president and provost] and not [the graduate student organization for a union]."

They proceeded to the sports page:

> "The caption on the photo misidentified the athlete on the squash court—he was from [neighboring school], not [our school]." "The sports calendar is wrong." The executive editor directed the editors to look at the bottom of page nine, saying, "the banner is kind of large telling people to recycle."

The executive editor said, "Let's take a look at [the arts and entertainment magazine that appears in the paper on Thursdays]." The editor of the magazine said, "What do you think of the cover? We had somebody else design it." The campus news editor said, "I liked the idea but I didn't like the text over his face." The editor of the article said, "the feature was well-researched but a little bland. And if anyone has beat reporters who want to write features, let me know." The former design editor who had since become a sports editor said, "four columns is too many to read across."

The executive editor said, "Ok, then let's hear about what's going on tomorrow." At this prompt, the executive editor called on editors around the room to each tell which stories were on the docket for the next day and how long they were. The final task of the meeting was to decide on the editorial written by the newspaper staff on the op-ed page. The topic was the organization of a graduate student union. The campus news editor said, "in light of the strike, maybe we should tone down the coverage." Another editor said, "even though it seems like an overload of [the grad student unionization group] coverage, there's no bigger news, and undergrads are ill-informed about it." The executive editor took a vote: "All in favor of running letters tomorrow. …" The group voted affirmatively.

Other Institutional Rules and Procedures

Reporters are expected to write stories at the paper's offices. On one visit, before the 6:30 p.m. editors' meeting, several students were sitting at workstations typing at computers. They intermittently looked down at their notes in notebooks, presumably from interviews and covering the event they were writing about, and one reporter listened to his tape recorder between sprees of typing. Later in the evening, this activity still took place. One female reporter who covered the university health system was typing at a computer. Her deadline had passed an hour prior. She told me that reporters were all supposed to come in to write their stories in the *DC* office for the paper the next day. As soon as the stories are submitted to editors, they get edited while the reporters are still at the office.

Although I observed one male staffer at the *DC* office listening to a tape recorder while typing up his article, which leads me to believe that the use of tape recorders is sanctioned and even encouraged, the reporter I accompanied to the live event did not use a tape recorder. Unlike the other two professional reporters at this event who wrote in their small, thin reporter's notebooks, the *DC* reporter had a large spiral notebook—the kind that is used to take notes in class. She wrote her notes feverishly in pencil, appearing to be taking the speaker's comments down word-for-word. When I asked her about this, she said that if she had asked, she believed that the event organizers probably wouldn't have allowed her to tape record (although one of the professional reporters at the event had used a tape recorder right in front of us). The *DC* reporter explained to me that she tries to err on the side of writing more than she thinks she'll need for her story. When her article appeared in the paper the next day on the front page as she had anticipated, it was full of selected direct quotes the speaker had made throughout the presentation, as well as quotes of student reactions that the reporter had obtained after the event concluded by approaching and interviewing audience members on the spot who were willing to speak with her. This reporter also said, "If I weren't lazy, I'd have tried to talk to [the speaker] himself."

Another institutional procedure that is followed is the use of wire services. Just as professional newspapers do, the managing editor told me that the *DC* subscribes to the Associated Press wire service and gets some of its photos from Getty. They get a "big discount for being a student news organization." The editors are responsible for choosing stories for the paper's world page from the AP wire.

Independence from the University

The paper operates completely independently from the university. It pays "100 percent" itself for rent and costs. At the time of my visits, the *DC* resided in a space in a building on the edge of campus, but it knew for some time that its lease would be expiring in the next few years and that the building would either be torn down or put to other uses. So the paper was thinking about buying or building a new place, but that would require $6 million, according to the managing editor. They had $1 million saved, so they would need to fundraise—an activity the editorial board, including the managing editor, executive editor, and some people from the business side were working on.

Cultivating and Working within Professional Culture: Comparisons to Other College Papers, Professional Newspapers or Journalism Societies That Set Norms and Give Awards

The student newspaper members have on several occasions displayed an awareness of their status as a college newspaper in relation to professional papers, other college papers with which they internally compete, and societies that recognize their work and that of their peers.

On my first visit to the *DC* offices, I noticed the many journalism awards hanging on the walls of the reception room. These included Pacemaker Awards, awards from the Association of Collegiate Press, and the College Newspaper Business and Advertising Managers, Inc. From my time spent with the sports editors, I found that they compared themselves to, and defended their own coverage against, other college newspapers. The sports editors made fun of [another college] paper for having only one sports story on each sport per week, while "the *DC* has tons more than that." One of the sports editors told me that they've received many awards for their coverage. I asked them if they also looked at professional papers, like the [major paper in the city], to which they pointed to three sets of newspaper pages pinned up on the wall of the office that displayed almost identical layouts. They explained that they had actually "stolen" a spread idea from the [city paper's] sports section in February, and that then, another college newspaper had copied them.

On one visit to the offices, the campus news editor told me that a reporter from the [city newspaper] had a grant during the semester so he didn't have to work, and instead attended classes at the university and helped the *DC* out each week. I never saw this [city newspaper] reporter during any of my visits.

In regard to the cultivation of professional culture, in the one particular situation of the *DC* public forum, it seemed that the executive editor was ill-prepared to deal with the attacks that were being made on the paper, and I wished I could have helped him out by offering some of the first amendment, freedom of speech and expression arguments that the press invokes. The executive editor had decided to hold this open forum for the university community and it was advertised in the paper. It was not an annual forum, and didn't take place the previous year, but had a few years ago, and the executive editor thought it was a good idea. Eight students were present in total, and the forum became centered around two issues, that while distinct, had overlapping elements.

The first issue was an anti-Palestinian ad that ran in the paper in the past week, and the second was an advertising insert that was in the paper that day, and was paid for by a pro-life organization, providing reasons why women should not have abortions or use birth control. Both issues centered on whether the *DC* should take these sort of paid advertisements at all, and if it decided to take them, what its policy should be on how the paid advertising content appeared. The executive editor fumbled around in trying to formulate defenses of the paper's policies.

At one point, he said, "You are right, we don't take sides, and we would run an ad on the other side the same way—we don't condone or endorse." But where he started to get off track was when he replied to a comment about the anti-Palestinian ad being racist by saying, "this is not libel as I understand it … I myself am of Palestinian descent. I had a personal problem with it, but I didn't get as strong an impression as you did about it."

The sports section's relationship with university athletic departments exemplifies the type of reporter-source relationships that scholars have discussed (Gans, 1980; Geiber, 1964; Sigal, 1973). The *DC* sports editors described having to go through the university press offices in the athletic departments at their home university and at other schools. They met with their home school athletic department every week to decide who they wanted to talk to and interview, and to discuss upcoming events. The senior sports editor told me how the press office could deny them access to coaches and athletes if they didn't like the angle the paper planned to take on a story, or if the paper had been critical of a team or its players in the past. "They can make life difficult," said the senior sports editor. However, since the student athletes of their home school are on campus and listed in the directory, the *DC* has all of their phone numbers, but cannot call them without the press office's permission for fear of "pissing them off," which could lead to the office blocking access in the future.

Informal Relationships within the Organization

In addition to formalized relationships within the newsroom, there were several types of informal interaction that took place between staffers. These relationships took the forms of inter-departmental rivalries, inter-office romances, friendships, and classmates.

The rivalry between the sports department and the news department was literally written on the wall: a computer-printed sign that hung on the wall in the sports office read: "NEWS SUCKS." The senior sports editor explained that there was indeed an inter-office rivalry that went on. He said that the news department thought that the sports department didn't do any work, but he believed they do. I asked the sports editor if sports was the only section that traveled to cover events. The editors said, basically, yes, although other sections could travel if they wanted to; in their eyes, this was clearly another fault of the news department.

The senior sports editor and the copy editor were roommates and joined the paper their freshman year together. Additionally, editors and their reporters are fellow students in the same university classes. When I asked if staffers ever socialize outside of the paper, the sports editors told me that very little time is left for that since they are always at the paper. I learned from the sports editors that staff members do date—the city news editor and another staffer were "going out." I was also informed that another couple had "hooked up" at a semi-formal the previous spring. Unlike in professional newsrooms, one episode at the *DC* implied that the place definitely served a dual purpose of workplace and social space. During one of my visits to the paper, while speaking to the editorial page editor in his office, he motioned to the little white fridge in his office and asked me if I wanted anything to drink—a beer? I declined, but he and the others proceeded to take them. A female student came in a little while later and popped one open too.

CONCLUSIONS ABOUT STUDENT NEWSMAKERS

From observations of, and discussions with, members of the *Daily Collegian* student newspaper, it is possible to see that this student journalism organization possesses many of the characteristics of a professional news organization, while also possessing traits unique to a specifically student-comprised college group. Some of the *DC*'s norms and routines were modeled after professional news organizations, while others, although they might resemble procedures in professional organizations, were created and learned by the members of the *DC* themselves in order to systematize and expedite necessary tasks.

The traits embodied by the members of the *DC* were age and context specific. They represent the transition from student and adolescent life to the adult and professional world and all of the struggles that accompany that. The reasons staffers exemplified and provided for their work at the paper, such as respect from peers, interest in what goes on around them, preparation for career, friendship, and sense of belonging, were all consistent with the stage of life they were in.

The atmosphere at the *Daily Collegian* is both professionally and socially constructed, and these two qualities of collegial socialization and aspiring professionalism are inevitably and unchangeably intertwined. The consequence of the tensions between this school culture and professional culture is that the *DC* falls short on some of the standards and ideals thought to be achieved by real news organizations, although one may wonder about this attribution to professional organizations as well. This study of the student journalists at this college newspaper has proven to be as much about "a student organization" as specifically a "news organization."

REFERENCES

Bivens, R. (2014). *Digital currents: How technology and the public are shaping TV news.* Toronto, Canada: University of Toronto Press.

Boczkowski, P. (2010). *News at work: Imitation in an age of information abundance.* Chicago, IL: University of Chicago Press.

Bodle, J. V. (1997). The instructional independence of daily student newspapers. *Journalism and Mass Communication Educator, 51*(4), 16–26.

Breed, W. (1955). Social control in the newsroom: A functional analysis. From *Social Forces, 33*, 326–355. In D. Berkowitz (Ed.), *Social meanings of news*, 1997: Thousand Oaks, CA: Sage Publications.

Fishman, M. (1980). The practice and politics of newswork. In *Manufacturing the news* (pp. 134–155). Austin, TX: University of Texas Press.

Gans, H. (1979). Values in the news and objectivity, values and ideology. In *Deciding what's news* (pp. 39–69 and 182–213). New York, NY: Pantheon.

Geiber, W. (1964). News is what newspapermen make it. In H. Tumber (Ed.), *News: A reader*, 1999, Oxford: Oxford University Press.

Gunter, B. (2000). *Media research methods: Measuring audiences, reactions and impact.* Thousand Oaks, CA: Sage Publications.

Klinenberg, E. (2003). *Media convergence: Time, space, and market pressures in high-tech news organizations* (Unpublished manuscript). New York, NY: New York University.

Marston, C. (1999). Learning to be a journalist: a feminist disability critique of cyborgs, college news-workers, and RSI work culture. *Journal of Communication Inquiry, 23*(2), 266–287.

McDevitt, M., Gassaway, B. M., & Perez, F. G. (2002). The making and unmaking of civic journalists; Influences of professional socialization. *Journalism & Mass Communication Quarterly, 79*(1), 87–100.

Sigal, L. V. (1973). Reporters and officials: The organization and politics of newsmaking. In H. Tumber (Ed.), N*ews: A reader*, 1999, Oxford: Oxford University Press.

Tuchman, G. (1978). *Making news: A study in the construction of reality.* New York, NY: The Free Press.

Usher, N. (2012a). Reshaping the public radio newsroom for the digital future. *Radio Journal, 11*(1), 65–79.

Usher, N. (2012b). Going web-first at the Christian Science Monitor: A three-part study of change. *International Journal of Communication, 6*, 1898–1917.

Watts, L., & Wernsman, R. (1997). Administrators as news sources for student journalists. *Journalism and Mass Communication Educator, 52*(1), 34–44.

Wearden, S. T., Hipsman, B. J., & Greenman, J. (1990). Racial diversity in the college newsroom. *Newspaper Research Journal, 11*(3), 80–95.

CHAPTER EIGHT

Laws, Rules AND Regulations

Student Media Learning Through Obstacles, Obstructionism and Opacity

SUSAN KIRKMAN ZAKE

When Chelsea Boozer was the editor of the University of Memphis' *The Daily Helmsman*, the campus police repeatedly harassed her after the paper reported on a campus rape in March 2012. Her troubles continued when an editorial criticized the police because they violated the Clery Act when they failed to notify students of the rape in a timely manner (Gregory, 2012). Two university administrators threatened to file student conduct charges against Boozer and police created false reports about her (Student Press Law Center, 2012). Then, in May, the student government attempted to disproportionately cut the paper's annual funding because, among other incidents, to cover the rape, Boozer pulled the reporter off a student government event. She routinely sparred with the university's legal department over public records and its misuse of The Family Educational Rights and Privacy Act (FERPA) to withhold information.

Boozer is an example of how those working as student journalists in training rightly or wrongly run into various walls and obstacles. Often, the difficulties are created by the misapplication or misinterpretation of laws, rules and regulations. Other times, the same laws, rules and regulations help to empower student journalists as they work to report information on their schools.

Boozer, who received the College Press Freedom Award in 2012 from the Student Press Law Center (Gregory, 2012) for the adversity she faced, uses the lessons she learned from her student media experience in her current job as the Little Rock City Hall reporter at the *Arkansas Democrat-Gazette*. She wrapped up a series of stories based on public records requests in early 2014 that led to the termination of the city's wastewater CEO. She's also battled with Little Rock's public housing chief over access to work orders and termination records (personal communication, December 29, 2014). When the chief tried to charge her $16,000 for the documents, she filed a "criminal complaint with our county's prosecuting attorney, who ultimately charged the housing chief with violating the Freedom of Information Act." The issue was to go to court in January 2015.

"In college," she said, "I thought it was unusual that I got so much push back from the student government types and some administrators due to my reporting on money or crime issues on campus" (C. Boozer, personal communication, December 29, 2014). She thought she was "dealing with a rare, unethical breed at the University of Memphis." But post-graduation, Boozer finds it's common to get push back when she is digging for facts and is thankful she had the college experience of "filing Freedom of Information Act requests and learning how to deal with those who don't comply with the state's open records law."

The exposure Boozer received from her battles helped her get her job in Little Rock—"every editor loves the fact that a young reporter has guts to take on her university's most senior administrators," she said (personal communication, December 29, 2014). And, she became "familiar with the techniques officials use to give reporters a hard time," and knows how to "stand my ground and be persistent in a quest for information" (personal communication, December 29, 2014).

A QUEST FOR OPEN GOVERNMENT AND TRANSPARENCY

Frank LoMonte, the executive director of the Student Press Law Center (SPLC), a nonprofit advocate for open government in schools and colleges, said many student media controversies "ebb and flow with the individual personalities at the publication. You get a class of students who are really, really fired up about getting a hold of records and investigating their school and then they graduate and things calm down again" (personal communication, November 17, 2014). LoMonte said the most frequent requests for help the SPLC gets from student media groups are about open government and transparency when dealing with colleges and universities—he hears the same complaint from the professional media when they deal with colleges. While private schools have much more leeway to restrict student expression and are largely exempt from public records laws, public schools, since they are government-funded, fall under rules and expectations that require openness and transparency. They must abide by the First Amendment—students working in media at state institutions have the constitutionally protected right to choose what they publish free from institutional censorship.

The universities don't think of themselves as providers of government services, but rather as businesses that are defensive of their reputations. "There is such a well-developed marketing apparatus at all of these institutions now that is obsessed with protecting the brand," LoMonte said. "And that's exactly the phrase they'll use. As if a college was a laundry detergent" (personal communication, November 17, 2014). He also points to diminishing investment in the schools by state government that leads to them thinking of themselves as "being answerable to their large donors and their corporate sponsors and not to the general public" (F. LoMonte, personal communication, November 17, 2014).

For students who go through experiences like Boozer's, LoMonte said there's "no question that once you have gone through the adversity of trying to pry information loose from a college that you're really well prepared to do that in the profession" (personal communication, November 17, 2014). Unfortunately, he said, the instinct of governments to resist disclosure continues off campus and government at all levels is obsessed with image protection—"they are all more secretive than they have ever been." Covering news in college, LoMonte said, is also good preparation for covering business, because with few exceptions, you can't "make a business tell you anything" (personal communication, November 17, 2014) so you have to cultivate sources and figure out alternative ways to come at your story.

The difficulties faced by students may build them into more tenacious journalists, but LoMonte is concerned there is an adverse effect:

> I think the adversity is really scaring people off substantive stories. Probably that top 10 percent of student journalists really rises to the challenge. But the other 90 percent just throws up their hands and says "the heck with it, I'll write that record review." (personal communication, November 17, 2014)

Colleges use a range of tactics to fend off requests for their records, including citing exemptions that "clearly shouldn't apply," LoMonte said (personal communication, November 17, 2014). "Many of them now stick you at the back of a very, very long line. Or, they'll come back with a $20,000 bill that they recognize is the functional equivalent of a denial." Several student newsrooms, including *The Daily Bruin* at the University of California, Los Angeles, publish "trackers" that keep tabs on the responsiveness of their schools to records requests (Records request, 2014).

Despite barriers, student journalists are finding smart and innovative ways to use public records, LoMonte said (personal communication, November 17, 2014). When they're successful, it's empowering. But the problem is there are "just some campuses that have such a culture of concealment that they don't respect their legal obligations." At the top of LoMonte's list is the University of North Carolina at Chapel Hill (personal communication, November 17, 2014). A long-running investigation into its football program, begun in 2010 after allegations of improper relationships with agents arose, led to the revelation in October 2014 that more than 3,000 students, many of them athletes, enrolled in classes they didn't have to show up for and received artificially inflated grades. The original 2010 investigation resulted in the NCAA issuing sanctions against the football program.

The University was sued three times by different media groups, including *The Daily Tar Heel* in November 2014, over the release of personnel records of nine employees facing disciplinary action in the academic scandal (Broward, 2014). The *Raleigh News and Observer* sued in January 2014 to get data shared with the university's accrediting agency and in 2010 (Doran, 2010) over the university's refusal to release records related to the football team and allegations of improper relationships with agents. The suit over the employee names was resolved in December 2014. The university cited FERPA in denying the requests.

OTHER BARRIERS TO INFORMATION

In addition to difficulty in getting access to public records, student journalists often struggle to get access to university officials for interviews without going through the university communications or public relations office. Many schools instruct employees to not speak with the media unless they've gained approval from the campus public relations office. The offices often request written questions in advance of an interview, or insist on email only interviews. LoMonte said the issue is rooted in the same "cultural problem that we have professional image makers who see themselves as gate keepers of the school's reputation." Athletic departments have controlled interviews for years, but now, LoMonte said, "it's getting to be like that for everybody on campus."

Mark Plenke, who advises *The Orion* at California State University, Chico and is the news professional in residence at Chico State, said one could argue with the public relations department that these policies have "no practical value" because it would be impossible for its staff to manage (Plenke, 2014). In a discussion on the College Media Association email listserv, Plenke asked, "Is the PR person going to script the response of the person your reporters would like to interview? Are all statements from everyone on campus going to be vetted and then issued by the PR department?" (Plenke, 2014, para. 5). Would the department make certain people "off limits" to reporters? And what would happen if "someone on the faculty, for example, violates the policy" and talks with the media without permission," he asked.

Closed presidential searches, where candidates never become publicly known until a new president is hired, are an increasingly common obstacle student media editors face in what may be the single most important issue they cover while in school. These searches are so pervasive the SPLC launched a blog called "Sack Secrecy" which says, "Unfortunately, too many colleges build their presidential selections around finding the most secret president, not the best one. Politically connected insiders win. Students lose" (Sack Secrecy, 2014, para. 3).

Andrea Gallo, editor in chief of *The Daily Reveille,* sued Louisiana State University in 2013 because it wouldn't turn over the names of the candidates for its vacant president's position. The university said it did not have the applications because the search was conducted through the school's private foundation, which hired an outside search firm (Moore, 2013). The experience taught Gallo, now a city hall reporter at *The Advocate* in Baton Rouge, "the importance of making a grand gesture" (personal communication, December 29, 2014). As a student journalist, she often felt she wasn't taken seriously or given the same level of respect as professionals. "That changed when I filed a public records lawsuit against my university. The clout of our college paper, *The Daily Reveille*, skyrocketed." University, government and community leaders were reading the paper's stories and it made Gallo "realize that being brave and asserting our watchdog role had paid off for us." According to the SPLC website, a letter from LSU denying the records request describes its efforts to maintain the candidates' privacy as a "standard" practice "in use by institutions of higher education across the country" and defends the action by maintaining that qualified candidates would not apply if the hiring process was public (Moore, 2013).

Gallo lost the first round of her lawsuit and eventually dropped it in favor of other media outlets that sued, including *The Advocate*. The latest appeals court ruling in December 2014 said LSU must turn over the names of four "finalists," rather than the broader pool of applicants (Kinchen, 2014). One of the lessons Gallo said she learned from the case is "that hot-button issues often come with a time limit, and that most people in the LSU community no longer care about whether these records are revealed" (personal communication, December 29, 2014).

Secret searches are not conducted at all institutions, LoMonte said, "but it tends to cluster at the highest and the public institutions" (personal communication, November 17, 2014). Some states, like Florida, Wisconsin, and West Virginia, have statutes that require searches to be conducted in the open. Some states require only finalists be made public, which can be worked around by naming only one finalist. But these are the exceptions and Florida's state legislature will consider legislation in 2015 (Ash, 2014) that would make searches for university presidents private. Wyoming's state legislature passed House Bill 223 in February 2013 and exempted "institutions of higher education from releasing identifying information in presidential searches if doing so is contrary to the public interest" (Nickerson, 2013, para. 5).

Michael Bragg, Editor-in-Chief of *The Appalachian* at Appalachian State University, ran a full, front page editorial requesting an open search when his school conducted a closed one for its new chancellor. At the time, Bragg said "We really wanted to shine a light on the issue for the public to see and for them to see why we think the closed search is not beneficial to the university" (McDermott, 2014, para. 9).

Daily Kent Stater Editor Daniel Moore (Our View, 2014) also ran a front-page editorial protesting the closed presidential search at Kent State University in April 2014, titled "The Worst Public Transparency Crisis Since the Last Time Around" referring to the previous presidential search at Kent State, which was also closed. While Kent State released some records relevant to the search process, it did not release the names of the candidates it considered before naming Beverly Warren, provost at Virginia Commonwealth University, as its new president. The university said those records resided with the private search firm it hired, which it also allowed to decide which records would be subject to public disclosure.

Some of the records revealed that the unnamed candidates were shuttled from the airport and snuck in loading dock entrances at the hotel to avoid the media waiting in the lobby. Search committee members also said their personal notes and documents were shredded by the university after they interviewed prospects (Biliczky, 2014). Kent State maintains it broke no public records laws. "Meetings were closed, they brought no candidates to campus, no one on the committee besides the chair would speak to us and as of today we still don't know any of the names of the people they considered, other than Beverly Warren" (Student Press Law Center, 2014, audio podcast). Moore said "This is $250,000 of public money and if there was any more prominent placement than front page, I would've done that too" (Student Press Law Center, 2014, audio podcast).

LoMonte (personal communication, November 18, 2014) said the rationale used by the universities is "a sitting president of a college will not compete for a job placing his current stature with his employer in jeopardy." But open searches lead to candidates who have more legitimacy and are a "better cultural fit" with the institution, he said. Open searches generate more public confidence and more diversity among the candidates, since the pool has to be shown to the public. Problematic closed searches at Penn State and the University of Wyoming, where the president hired during a closed search left after only four months, lead LoMonte to ask, "How have people not been burned by this enough times to recognize the cost of secrecy?" (personal communication, November 18, 2014).

FERPA, DISCIPLINARY BOARDS, AND PARKING TICKETS

The Family Educational Rights and Privacy Act, enacted in 1974 by President Gerald R. Ford, is intended to protect students and parents, and limits the education records that can be given out by educational agencies or institutions (Family Educational Rights, 2014). But it is often cited and frequently misunderstood and misinterpreted when universities withhold records from student journalists.

Some recent examples:

- Student government election results were withheld for eight months from editors of the *Ferris State Torch* (Wittman, 2013).

- *The Oklahoma Daily* filed a lawsuit in May 2013 against Oklahoma University after the school refused to release parking ticket information to *Daily* staff member Joey Stipek. OU president David Boren ordered the release of the data in November 2014; he said "In my opinion, the records in question are traffic violation records, and are not the kind of sensitive student records, which are covered by the Federal Educational Rights and Privacy Act" (Burris, 2014).
- The University of Michigan repeatedly cited FERPA when it refused to discuss the expulsion in January 2014 of its football team's starting kicker for sexual misconduct that was originally reported in 2009. Records of the outcome of disciplinary proceedings involving crimes of violence, including sexual assaults, are exempt from FERPA (Goldstein, 2014).
- A public relations spokesperson at Columbia University in New York City cited FERPA when he told the student newspaper they couldn't write about graffiti in a toilet stall. The writing contained the names of four young men suspected of being campus rapists (F. LoMonte, personal communication, November 18, 2014).

"There is this cult of FERPA that has developed among college attorneys that is really of the zeal and fervor of a religion," LoMonte said. "You hear these attorneys who are otherwise educated people swear to you that Congress meant to make college students invisible" (personal communication, November 18, 2014). It doesn't help that FERPA is "a terribly written statute," and can be interpreted "so that every dealing a student has with his institution is secret," LoMonte said. The Supreme Court has made clear that "a document is not protected by FERPA unless it is both directly related to a student and centrally maintained in a file corresponding to that student." FERPA is, in fact, a "very narrow privacy statute that the college attorney community has succeeded in transforming into a blanket of secrecy" (F. LoMonte, personal communication, November 18, 2014).

FERPA is especially contentious when it's related to the actions of campus disciplinary boards. In 2014, the Department of Education was investigating 55 colleges for potential Title IX violations linked to the improper handling of campus sexual assault cases, often by campus disciplinary boards. Title IX prohibits discrimination on the basis of sex in education programs or activities that receive federal funds. LoMonte said student conduct boards were intended to handle "really minor acts like sneaking a beer into your dorm." But now, he said, they're being asked to "adjudicate crimes that anywhere else except on the campus of a college would be felonies carrying 20 years to life in prison" (personal communication, November 18, 2014).

The boards were built to give a clean slate and a second chance to young people for minor, petty offenses that don't harm anyone. "Nobody minds if you give a second chance to a kid who smoked one puff of pot but nobody gets one free rape" (LoMonte, personal communication, November 18, 2014). The fact that disciplinary boards are lumping serious crimes like rape into the same category with underage drinking and don't follow the stringent due process of a courtroom has led to increased tension over their secret operations. Another complaint against student conduct boards occurs when they're used to threaten or punish student journalists for doing their jobs.

The Foundation for Individual Rights in Education said in its Guide to Due Process Procedure on Campus that college students "run a significant risk of being found responsible for a minor or, indeed, serious offense even if [they] are innocent" because "campus courts lack the kinds of basic fact-finding mechanisms and procedural safeguards that a decent society should provide" (Silverglate and Gewolb, 2013, Part 1: Introduction section, para. 3).

At the University of Virginia in 2011, student judiciary charges were ultimately dropped against Jason Ally, Editor-in-Chief of the *Cavalier Daily,* after he and four other managing board members wrote an editorial disclosing a reporter had plagiarized material in multiple articles (Velz, 2011). The charges, filed through the University Judiciary Committee, were for breaching the confidentiality of the writer's disciplinary case.

Dylan Bouscher, editor-in-chief of the *University Press* at Florida Atlantic University, faced student conduct charges in 2013 after university police said he refused to leave when told he was photographing within the boundaries of a crime scene (Sunne, 2013). Bouscher said he was in a public space. In cases like Bouscher's, LoMonte said, "the imposition of disciplinary sanctions by college 'kangaroo courts' can impose real and lasting chilling effects on journalism" (personal communication, November 18, 2014). Newsrooms are workplaces, he said, and "it is never under any circumstances appropriate for a campus disciplinary board to invoke its authority to punish workplace behavior" (LoMonte, 2013).

CONCLUSION

LoMonte remembers a time when every Monday at the University of Florida, the president held a senior staff meeting that included all the vice presidents and associate vice presidents. They sat around a conference table and the media would come in. "You would sit there with your note pad and you would get story idea after story idea." He said it was a climate of "we don't keep secrets here. We're answerable to the public" (F. LoMonte, personal communication, November 17, 2014). For many student journalists, the situations they deal with are far less transparent, and in some cases, openly adversarial. But most agree there are also lessons attached that prepare them for their professional careers.

Moore, the *Daily Kent Stater* editor who is now an energy reporter for the *Pittsburgh Post-Gazette,* realized that in the face of obstructionism, he should not "become bogged down with the long-term fight for transparency. There are just-as-important quick stories to be done on the secrecy itself and how it can throw what should be an easy, open and exciting democratic process through the wringer" (personal communication, January 4, 2015). Boozer views her struggles at Memphis as "ironically lucky" and said she "wouldn't trade my experience in college—no matter how challenging to get through at times—for the world" (personal communication, December 29, 2014). People sometimes recognize Gallo as "the girl who sued LSU." She said she is "more than happy" for that to be her current legacy, but hopes to "use those lessons and experiences to do even more important work" in her professional life (personal communication, December 29, 2014).

LoMonte sees student journalists like these figuring out creative ways to use the power of Freedom of Information laws. For him, the quest for transparency boils down to the public's right "to know if the people running campus safety or running the business office or running academic affairs know what they're doing. To build a wall of insulation around them is really disrespectful of public accountability" (personal communication, November 17, 2014).

REFERENCES

Ash, J. (2014, December 16). *Bill to close presidential searches refiled for 2015 legislative session.* Retrieved December 26, 2014, from http://news.wfsu.org/post/bill-close-presidential-searches-refiled-2015-legislative-session

Biliczky, C. (2014, April 12). *Kent State shredded documents to hide information about presidential search, committee members say.* Retrieved December 25, 2014, from http://www.ohio.com/news/local/kent-state-shredded-documents-to-hide-information-about-presidential-search-committee-members-say-1.480333

Browder, C. (2014, November 24). *Refusal to disclose names of those disciplined after UNC-CH fraud probe prompts lawsuit* [Radio broadcast]. Retrieved December 28, 2014, from http://www.wral.com/refusal-to-disclose-names-of-those-disciplined-after-unc-ch-fraud-probe-prompts-lawsuit/14209317/

Burris, P. (2014, November 12). *Boren responds to lawsuit, orders release of parking ticket information.* Retrieved December 11, 2014, from http://www.oudaily.com/news/boren-responds-to-lawsuit-orders-release-of-parking-ticket-information/article_37394e14-6ab4-11e4-b657-a398fb496feb.html

Doran, W. (2010, October 29). *Eight suing UNC officials for records.* Retrieved December 22, 2014, from http://www.dailytarheel.com/article/2010/10/eight_suing_unc_ocials_for_records

Family Educational Rights and Privacy Act (FERPA). (n.d.). Retrieved December 27, 2014, from http://www2.ed.gov/policy/gen/guid/fpco/ferpa/index.html

Goldstein, A. (2014, February 1). *University of Michigan says it cannot comment on football player expelled for sexual misconduct because of FERPA.* Retrieved December 23, 2014, from http://ferpafact.tumblr.com/post/75073849910/university-of-michigan-says-it-cannot-comment-on

Gregory, S. (2012, November 6). *Memphis journalist and College Press Freedom Award winner tells her story.* Retrieved October 29, 2014, from http://www.splc.org/blog/splc/2012/11/memphis-journalist-and-college-press-freedom-award-winner-tells-her-story

Kinchen, H. (2014, December 30). *LSU wins partial reversal in lawsuit over presidential search documents.* Retrieved December 30, 2014, from http://theadvocate.com/news/11213914-123/lsu-wins-partial-reversal-in

LoMonte, F. (2013, October 10). *Should reporter who lied be named by her paper?* Retrieved from http://www.collegemedia.org/search/results/?q=media+association+electronic+mailing+list

McDermott, C. (2014, January 17). *Q&A: Appalachian editor calls for open chancellor search in front-page editorial.* Retrieved December 26, 2014, from http://www.splc.org/blog/splc/2014/01/qa-appalachian-editor-calls-for-open-chancellor-search-in-front-page-editorial

Moore, D. (2013, April 1). *LSU student editor files suit against school seeking names of presidential candidates.* Retrieved November 21, 2014, from http://www.splc.org/article/2013/04/lsu-student-editor-files-suit-against-school-seeking-names-of-presidential-candidates

Nickerson, G. (2013, February 5). *Point and counterpoint: Closed UWYO presidential search process.* Retrieved December 28, 2014, from http://wyofile.com/gregory_nickerson/20080

Our View: The worst public transparency crisis since the last time around. (2014, April 20). Retrieved December 27, 2014, from http://www.kentwired.com/opinion/article_ae7cf8fa-c8e2-11e3-8889-0017a43b2370.html

Plenke, M. (2014, January 31). *Need advice dealing with campus PR Dept.* Retrieved from http://www.collegemedia.org/search/results/?q=media+association+electronic+mailing+list

Records requests. (n.d.). In Dailybruin.com. Retrieved December 28, 2014, from http://dailybruin.com/records-requests/

Sack Secrecy. (n.d.). *Introduction.* Retrieved November 24, 2014, from http://sacksecrecy.tumblr.com

Silverglate, H., & Gewolb, J. (2013, December 7). *Guide to due process procedure on campus.* Retrieved December 29, 2014, from http://www.thefire.org/pdfs/due-process.pdf

Student Press Law Center. (2012, June 12). *2012 College press freedom award winner Chelsea Boozer* [Video file]. Retrieved from http://vimeo.com/52899774

Student Press Law Center. (2014, April 28). *In North Carolina and Ohio, two student editors discuss secret presidential searches* [Audio Podcast]. Retrieved December 24, 2014, from http://www.splc.org/multimedia/185

Sunne, S. (2013, September 26). *Florida college punishes student reporter for failing to leave suicide scene.* Retrieved December 27, 2014, from http://www.splc.org/article/2013/09/florida-college-punishes-student-reporter-for-failing-to-leave-suicide-scene?id=2611

Velz, P. (2011, October 11). *Questions linger after charges dropped against UVA editor.* Retrieved December 27, 2014, from http://www.splc.org/article/2011/10/questions-linger-after-charges-dropped-against-uva-editor

Wittman, A. (2013, December 4). *Maddening middleman—The Ferris State Torch.* Retrieved December 27, 2014, from http://fsutorch.com/2013/12/04/maddening-middleman/#comment-3978

CHAPTER NINE

The Learning Curve at Student Media

Old Challenges and New Trends

HAIJING TU

Experiential learning has been the standard procedure in journalism education for over 100 years despite the divide between theory and practice in the field (Brandon, 2002, Greenburg, 2007). According to Brandon (2002), experiential learning is a process during which a person experiences an event, acquires competencies, and then compares the knowledge gained with knowledge obtained in similar situations. The core characteristic of experiential learning, according to Moore (2010) is students' direct engagement in productive work outside the classroom. As a crucial element of higher education, experiential learning is expected to ease the transition from school to a career for students (Moore, 2010).

In the United States, student media organizations offer unique experiential learning opportunities for students interested in working in professional media. Beyond that, student media play a critical role in shaping the practice of professional journalism (Schmidt, 2014). Across the nation, tens of thousands of students participate in the daily operation of student media on college campuses. Therefore the learning process in student media should be of great interest to educators as well as media professionals.

As digital media quickly evolved by offering unprecedented levels of interactivity and accessibility, the norms of production have changed from traditional mass media to web-based media. As a result, different expectations are required of student media organizations. The skills and values of journalism that are sought in new graduates are now different from what was asked by traditional media

(Cremedas & Lysak, 2011; Pew Research Center, 2008). In addition, the media landscape is filled with uncertainty with respect to the future of both old and new media. The situation has been made even more difficult as current students have been reported to "grow less interested in the news, regardless of how it's delivered" (Rogers, n.d.). Together, these factors pose challenges to the learning effectiveness of student media organizations.

This chapter discusses the typical learning curve in student media today and explores the role of experiential education in journalism in student media in the digital age. To address persistent challenges faced by faculty, staff, and students in accomplishing what is expected of them by the real world, the author studied the student media organization at Indiana State University (ISU), a Midwest public university with an enrollment of about 12,000. This was accomplished by conducting in-depth interviews with the student media director, managers, and student workers. Similar to many other public universities in the United States, student media organizations at ISU operate their own radio stations, one student newspaper, and one campus TV channel.

THE STORY OF STUDENT MEDIA ORGANIZATIONS AT ISU

Staffed mostly by students, student media organizations are active agents in creating and promoting college campus cultures. At ISU, students have autonomy in the production and distribution of student media with minimum supervision of staff members. Student workers are mostly recruited and hired by their peers at ISU. These organizations offer volunteer, hourly paid positions, and freelance opportunities for those who are interested in gaining media work experiences. Recently, the student media at ISU has had a digital team working on web content in co-operation with radio, video, and the student newspaper. There are four full time staff members and all of them have had experiences working in the industry before working for student media. Including volunteers, there are about 125 students working for student media at ISU of which approximately 50 are hourly student workers.

In August 2014, Indiana State University acquired WMHD from the nearby Rose-Hulman Institute of Technology after FCC approval (Taylor, 2014). Since then the radio station has established collaboration with WFYI public media to stream National Public Radio public affairs and news programming to west-central Indiana. Ultimately, the acquisition expanded opportunities for student broadcasters who are still interested in traditional media despite the fragmentation of media platforms.

Faculty and staff members and a few student workers at ISU student media discussed learning objectives and learning effectiveness at student media in the digital age during one-on-one interviews with the author. They revealed their thoughts about the changes they have experienced and discussed challenges and new trends in students' learning experiences in student media.

THE PAST OF STUDENT MEDIA AND EXPERIENTIAL LEARNING

Student Media in the Digital Age

The media industry experienced change in the transitions from print to radio and from radio to TV. This can be seen as parallel to what has happened in the transition from traditional broadcasting and print to digital media. In such comparisons, newer forms of media are simply considered new containers for an older message (McLuhan, 1964). During the transition from traditional media to new media, instead of studying "old against new," Geiger and Lampinen (2014) were more interested in studying the coming-of-age of media. They noted that this media "does not involve the new replacing the old, but rather the emergence of complex assemblages that are produced when the new is incorporated into the old, or vice-versa" (p. 334).

Student media organizations are no exception to the impact brought by new media. Cremedas and Lysak's (2011) research indicated that the web has found its place as a primary news presentation tool in local television newsrooms, and news directors "overwhelmingly prefer to hire people with existing web skills" (p. 41). According to Cremeda and Lysak, the increasing emphasis on web site content has prompted many broadcast journalism educators to reevaluate and change curriculum and program focus to ensure students acquire the necessary web skills.

In his survey research conducted in 2011, Schmidt (2014) found that students reported feeling marginally prepared to work professionally as journalists. In particular, they reported that students lacked skills related to new media production. In addition, Adams (2008) found that preparedness to practice technology-related skills was reported to be low among college students majoring in print journalism.

A technology driven media landscape also has disadvantages. There was a time that student media had to struggle with the limitations of equipment (Buckingham, Harvey, & Sefton-Green, 1999). With the Internet and spread of personal multimedia technology, these difficulties have been mostly ameliorated. However, according to Buckingham et al. (1999), the availability of new media technology "does not necessarily make the act of constructing an image anymore conscious or deliberate" (p. 16). This argument was echoed by a 2008 Pew Research Report on the challenge of the web to journalistic values. Pew found that many journalists saw the weakening of journalistic values as a result of the pressure for journalists to quickly develop these new digital skills (Pew Research Center, 2008).

Student Media and Professional Media Practice

Journalism education used to serve the purpose of training students to find information, shape it into an accurate story, and transmit it as quickly as possible to a mass audience via a mass medium (Carey, 2000). Carey noted that "professional education … has been driven by the desire to have a workforce that is moral, orderly, habitual, and conservative" (p. 16). But the purpose of traditional journalism education is now being questioned as the media industry moves from the one-way communication model to an interactive one.

According to Mensing (2010) the transmission-driven, industry-conceived model of journalism is now out of date. Mensing noted three major consequences of the industry-centered patterns of journalism education: (1) a focus on creating professionals, despite the trends toward de-professionalization; (2) teaching that reinforces one-way communication; and (3) a focus on socializing students for the newsroom more than engaging in critical inquiry. Clearly, teaching communication as a process of transmission from producer to receiver is not going to help students handle the work in an instantly interactive media industry. As a remedy, Mensing suggested that the focus of journalism education should be moved from the industry to the community in order to better apply new forms of news creation, production, editing, and distribution.

Among the trends toward de-professionalization, plagiarism is among the top concerns for journalism practice when everything is available to select, copy, and paste. Beyond that, another danger is that students have started conducting interviews via email only, which is considered a dangerous practice because the lack of human contact (Greenwald, 2004). It is not surprising that relying on limited sources is also a common disadvantage in student media practice (Schmidt, 2014).

RESEARCH QUESTIONS

Considering that the digital/social media revolution is making the one-way communication model obsolete, media professionals are expected to continually pick up new skills (Geiger & Lampinen, 2014). Student media are no exception when adapting to new media. In order to accomplish effective learning outcomes in student media organizations and embrace new norms and practices brought by digital media, educators and professionals should be informed about the current learning experience at student media and its impact on students.

The research questions of this study follow:

1. What are the learning outcomes that are currently being set in student media organizations?

2. Is there a gap between training at student media organizations and the expectations of professional media practitioners?
3. What are the recent trends addressing new media usage in student media organizations?

For this study, five faculty and staff members affiliated with or who previously worked for student media at ISU were interviewed, as well as four student workers. The interviews were conducted in the fall semester 2014. Consent was gained through email confirmation. Participants were asked the following questions during the semi-structured interviews:

1. What are the differences between student media and professional media institutions?
2. What are the objectives of experiential learning at student media organizations?
3. What has been the biggest challenge for students in accomplishing these learning goals?
4. Are there modes of assessing the learning outcomes of student media organizations?
5. The traditional media industry is transforming under the impact of new media, what impact does it have on student media?
6. What are the changes in the audience for student media?

All the interviews were audio taped and were transcribed later. Answers to each question were analyzed by comparison and categorization. The lively discussions reflected the common issues concerning today's campus media, especially with respect to the acquisition of skills in new media technology and social media.

FINDINGS ON OLD CHALLENGES AND NEW TRENDS

Old Challenge No. 1: Learning Outcomes at Student Media

Because experiential learning was considered extracurricular, student media at ISU did not have a curriculum set up for student workers. Learning at student media was not included in teaching and learning assessments by the University. But staff members at student media do teach regular media classes in the Communication Department nearby, and these classes were evaluated regularly. As a faculty member who used to work at ISU student media, L. Hensen (personal communication, October 6, 2014) believed that involvement in student media organizations certainly benefited students in their knowledge and skills:

> I think student media has an opportunity that we lack in the classroom of experimentation, you know, I think by virtue of how course syllabi are created and set up, it is difficult to include experimentation in a course setting. You want there to be definitive exercises they can master and move on. So student media is especially important because it involves opportunity for them to play and try a type of software or new approach to something with very low stakes.

This kind of experimentation makes working at student media organizations a learning experience that students are not able to obtain elsewhere. For example, participants in this study believed that students had the opportunity to accomplish civic engagement with the community by working at student media. According to L. Hensen (personal communication, October 6, 2014) students were more likely to learn from the various topics they covered in their news assignments:

> I don't know a better tool for increasing students' overall knowledge base than student media. I think it's the most effective tool for helping students become civically engaged. They certainly learn a great deal about politics, government, and business, and certainly students who work with ad sales get the business side of the organization … The whole range of topics you cover in a paper, crime, and arts, and culture, and government, and on and on. The students have to learn about those things because they are the ones covering it. I mean what other organizations are there that allow you to do that?

The assessment of learning effectiveness at student media was not formal, and was often performance based. After being screened for their credentials during the recruiting process, selected candidates were supervised by student directors in each department of ISU student media organizations. R. Green, the manager of the student radio station talked about supervising student workers (personal communication, August 28, 2014):

> We meet once a week. I meet with directors to see what their department is up to, where they are at. Obviously we are not going too hard on them because they are still students they are still learning, but there is definitely some accountability, … There are certainly some people who test patience … If you care, then they care.

The faculty and staff working with student media organizations measured the success of learning experiences by using multiple criteria. First, they looked at how helpful the experience was for students' career paths. Such indicators include, but are not limited to, the skills they acquired that allow them to function in another context and the chance they were able to secure internships, scholarships and full time jobs. The director of student video production specifically talked about how proud he felt about seeing his students go on to work in broadcast, TV, and film production (J. Perkins, personal communication, September 5, 2014). For students who chose not to work in the media industry, J. Perkins believed that the experience working in student media organizations was still valuable in teaching them how to be on time, dependable, and work with others. Secondly, according to P. Glende, the executive director of ISU Student Media, the success of student media in offering such experiences was also measured by students' participation (personal communication, August 25, 2014):

> We can assess our success by whether they [students] return to us because we ask people to come in as volunteers and work for us, and if they come back to us over and over again, we think we are successful. Their continuing participation is a measurement of our success, that our staff is growing, more students are becoming committed enough to engage in the discipline of producing media. Participation itself is success.

Students offered positive feedback based on their learning experiences at student media at ISU. They typically dedicated 12 hours per week working in student media organizations. They noted that their student media experience was a good resume builder. A student was happy to move from a rural area to the city and get to work with people from "all over the place," because "that's something that helps you to adapt to the real world from different people" (J. Lafary, personal communication, September 12, 2014). Another student, R. Richards, was happy about learning the skills on his position in the radio station (personal communication, September 12, 2014):

> I think I learned pretty much everything I wanted to. I dealt with the mixer, how to work with the music part. The only thing … maybe to set remote things like that, to work with off campus stuff. As a deejay that's a part I would like to learn more. That's the only thing I'd like to get.

Overall, student media organizations are where students learn to be professional and get prepared for the real world. Learning effectiveness will eventually be tested by that real world. P. Glende (personal communication, August 25, 2014) addressed the assessment of learning outcomes of student media:

> A lot of students who came in and work for us for the most part aren't interested in learning life skills, they are interested in having fun. They are interested in participating in media. But we know, as educators, that it is important experience for them, it's potentially life-shaping. You have no idea where it might lead you, or how you might use the experience at student media.

Old Challenge No. 2: Student Media vs. Professional Media

Professional print media have seen a decline in readership since the beginning of the millennium. Researchers found the percentage of daily print newspaper readers in the U.S. had fallen nearly 20 percent since 2001 (Sebastian, 2013). The same pressure is being felt by student media organizations at ISU. Although there was no concrete data, one interviewee with a lot of experience working in both the industry and student media admitted that student media are not as popular as they were 20–30 years ago (P. Glende, personal communication, August 25, 2014). Glende said:

> My sense is that there's been a decline in the interest in student media ... This is especially true at our newspaper. We understand that newspaper readership is declining, newspaper relevancy is declining, and that's reflected in interest in our enterprise ... There was a different time, before CNN, the 24/7 phenomenon. At that time, all newspapers had higher standing, higher readership. Readership was already in decline but it was not an issue at that time, because circulation was still very high, radio station participation was still very strong at that time, with no such thing as even cable television at that time.

Given the decline of readership, student media organizations gave priority to producing good content. P. Glende expected that the focus on content would help to slow down such a decline and attract students to student media in the digital age (personal communication, August 25, 2014):

> We are doing several things to make that happen, we think there is still room for the role of audiences, even though we acknowledge challenges all along. Digitally engaged students who may not want to pick up a newspaper and read it. We know it. We are working on several web initiatives that we think are important to us. And we think that a free general circulation paper is still a viable way of telling stories about students and for students. I personally believe that it is still all about content, about telling good stories, about representing communities fully, informing communities and giving them useful information in an engaging form.

Of all four student media staff members the author interviewed, three of them had prior experience working in professional news agencies. They were conscious about the differences in standards of success between professional media and student media. "The way it is different from the real world is that you are working more with the students," said R. Green (personal communication, September 12, 2014) "when working in commercial radio, you have to follow more standards, student media has more freedom, be yourself a little more."

For students who were aiming at working for professional media institutions, they also face challenges to fulfill their experiential learning goals. Among them, time management turned out to be the top challenge for student workers:

> Without a doubt, I would say that the most significant challenge [for student workers] is time management. I don't mean to be critical to students in saying that. By time management, I mean, they are in class, they are working outside of class, many of them are working off campus, they may have family commitments at home, they have rigorous schedules, they have a social life, involved in Greek activities, my impression is time consuming activity. They just ran out of time to do this when we need them to do it reliably and predictably. The most difficult issue is to have them here, having the time, making time to do this. (P. Glende, personal communication, August 25, 2014)

Also, the lack of courses in management in their major became a concern for students in preparing fully for their future careers. R. Green recalled experiences at another institution he was familiar with and discussed possible changes in curriculum that could benefit students (personal communication, August 28, 2014):

> They still have Radio and TV majors, but they've also moved towards the promotion of the media and management side, so they are getting more business experience as well as their TV, radio, Photoshop experience as well. The lack of management courses and the lack of promotion classes is one of the things that we should work toward.

NEW TRENDS IN STUDENT MEDIA BROUGHT BY NEW MEDIA

The convenience of technology has posed new challenges to experiential learning in student media. Interviewees observed that campus discourse had been occurring already on multiple online platforms. Student media were experiencing the transformative force brought by new media. P. Glende (personal communication, August 25, 2014) described the pressure from new media:

> There most certainly is pressure, I don't know if we have necessarily felt it. There are not people out there telling us that we should do more online. But we know we should. The pressure is internal. We are not feeling pressure, we are exerting pressure. We know that giving students experience in digital media is important.

A team of ISU students interested in digital media was formed. They created a website and began preparing to get involved in all aspects of the operation of the student media organizations. C. Clark, the manager of student media web content, was leading the digital team, and expressed a positive outlook for the future of this area, believing that student media will rely more on the digital team. According to C. Clark (personal communication, September 3, 2014), their daily tasks included "digitalizing the content produced by the newspaper, the radio, and the Sycamore video, building cohesion with logos and banners, etc." As a result, graphic design and coding became desired skills for student workers in student media organizations.

Students now have to learn to multitask at student media. According to the interviewees, student workers in the past would pick one area they were interested in and focus on that one area when working at student media. For example, students used to pick from radio, TV, or the newspaper and focus on it. All areas used to be divided. But today's students are striving to be able to work in each of the different fields. "Print journalism goes out. When students write a story they also need to take a video," said J. Perkins (personal communication, September 5, 2014).

Social media is an important part of this new trend. C. Clark (personal communication, September 3, 2014) explained:

> One of the things we are happy with is how we have taken social media. And we've taken it from a toy of novelty to utilizing that as part of our digitalized efforts. Do we have a full swing digital media policy in place? No we don't. But we are headed to that direction. Just figuring out a tone, a theme, and a voice for each side of the tweets is huge accomplishment that I think the digital team has been at the forefront of.

With multiple platforms of information distribution came large amounts of information for students to absorb. As citizen journalism knocked on the screens of their phones, student media organizations looked for ways to reach their fellow students. One interviewee believed in the value of "help[ing] students assess, capture, and analyze what's important from the flood of conversation in their community" (L. Hensen, personal communication, October 5, 2014). It also explained the necessity for student media to move online and compete for attention from students.

CONCLUSION

This chapter has explored major challenges brought by new media on the experiential learning process at student media using the example of a Midwest public university. The interviews with student media faculty, staff, and student workers revealed many aspects of the operation of a student media organization. The implications of this study should be useful for scholars and professionals interested in the transition of student media under the impact of new media.

"Journalism is something other than media and communication. It is a particular kind of democratic practice" (Carey, 2000, p. 22). It came to the attention of the author that citizen journalism and the wonders of technology working together had changed the playing field of student media. Given the decline of audience size for student media in the past two decades, the experiential learning experience at student media may see a new direction in engaging students with local community, rather than simply offering information based on the one-way transmission model.

Either by being independent or by being forced, student media have to cope with the rise of new media/social media. The gap between professional media and student media in terms of techniques required to fully function at work will eventually close. Content is still considered the most importance competency of student media. Skills in new media technology will follow closely, and it's a good start that students are now receiving training in using these technologies at student media.

Buckingham et al. (1999) noted that "[t]he impact of technology can make depends on the social contexts and relationship into which it enters" (p. 20). The focus of student media is campus life, which appeals to the audience with relativity, proximity, and timeliness. Now social media take the place and offer more accessibility at students' fingertips. For student workers at student media to expand the influence of their work, they have no choice but to join the competition for attention using new media. This is also going to be part of their job responsibilities after entering the professional media industry.

The last implication of this study can be instructive to curriculum designers in journalism and media studies. Student media are often closely tied to the department of communication and journalism in a university. The rise of social media and new media has been injecting social media studies into the curriculum of academic departments while traditional news courses stagnate. The give and take in this process sometimes involves tough decisions, since the learning outcomes for students will be weighed against the established institutional structure.

For future research, scholars may want to observe this phenomenon from an alumni's perspective. Alumni who work in professional media will have to account for the learning effectiveness of their experiences in student media. To find solutions to these challenges, more interviews with university administrators should be conducted in order to draw a larger picture of the position of student media in higher education.

REFERENCES

Adams, J. W. (2008). Industry guidance could help J-programs prepare print majors for convergence. *Newspaper Research Journal, 29*(4), 81–88.

Brandon, W. (2002). Experiential learning: A new research path to the study of journalism education. *Journalism and Mass Communication Educator, 57*(1), 59–66.

Buckingham, D., Harvey, I., & Sefton-Green, J. (1999). The different is digital? Digital technology and student media production. *Convergence: The International Journal of Research into New Media Technologies, 5*(10), 10–20.

Carey, J. (2000). Some personal notes on US journalism education. *Journalism: Theory, Practice, and Criticism, 1*(1), 12–23.

Cremedas, M., & Lysak, S. (2011). "New media" skills competency expected of TV reporters and producers: A survey. *Electronic News, 5*(1), 41–59.

Geiger, R. S., & Lampinen, A. (2014). Old against new, or a coming of age? Broadcasting in an era of electronic media. *Journal of Broadcasting & Electronic Media, 58*(3), 333–341.

Greenburg, S. (2007). Theory and practice in journalism education. *Journal of Media Practice, 8*(3), 289–303.

Greenwald, M. (2004, August). Coping with the challenges technology brings. *Quill Magazine, 92*(6), 25.

McLuhan, M. (1964). *Understanding media: Then extension of man.* New York, NY: McGraw-Hill.

Mensing, D. (2010). Rethinking [again] the future of journalism education. *Journalism Studies, 11*(4), 511–523.

Moore, D. T. (2010). Forms and issues in experiential learning. *New Directions for Teaching & Learning, 124,* 3–13.

Pew Research Center. (2008). *The Web: Alarming, appealing and a challenge to journalistic values.* Retrieved from http://www.stateofthemedia.org/files/2011/01/Journalist-report-2008.pdf
Rogers, T. (n.d.). *The technology of journalism improves, but young people still ignore the news.* Retrieved from http://journalism.about.com/od/trends/a/youngpeople.htm
Schmidt, H. C. (2014). Training tomorrow's journalists: An analysis of student newspaper content. *Pennsylvania Communication Annual, 70*, 11–40.
Sebastian, M. (2013, June 10). The best (and worst) cities for newspapers. *Advertising Age, 84*(24), 3.
Taylor, D. (2014, August 20). Indiana State acquires Rose-Hulman radio station, will add NPR programming. *ISU Newsroom.* Retrieved from http://www2.indstate.edu/news/news.php?newsid=4052

CHAPTER TEN

Dealing WITH Disruption

Learning Industry Lessons on the Job in Student Media

SUSAN KIRKMAN ZAKE

Student newsrooms today find themselves "mired in the innovator's dilemma" alongside their professional colleagues as they watch traditional sources of revenue fall and are forced to make "a false choice between today's revenues and tomorrow's digital promise" (Christensen, Skok, & Allsworth, 2012, para. 2). Newsrooms, and particularly newspapers, are experiencing a phenomenon called "disruptive innovation," where "businesses grow, become successful, and falter as nimble start-ups muscle in on their customers" (Christensen et al., 2012, para. 1).

The theory of disruptive innovation was developed by Harvard Business School Professor Clayton M. Christensen in *The Innovator's Dilemma* (1997). In the case of newsrooms, the start-ups take the form of "new-market disruptors" and deliver products over the Internet that are "faster and more personalized than that provided by the bigger, more established news organizations" (Christensen et al., 2012, para. 12). The new businesses start at the low end of the existing systems, entering with a lower cost and lower profit margin formula—they move up the "value network" and eventually siphon off customer base (Christensen et al., 2012, para. 10). Applied to journalistic institutions, disruptive innovation plays out as a search for new business models to close the revenue gap between print and digital and allows traditional news organizations to remain relevant to their audience.

In its *State of the News Media Report* (Mitchell, 2014), The Pew Research Center found some reasons for optimism for the survival of the news media, since disruptors like *BuzzFeed* and *Mashable*, built around an "innate understanding of technology" (Mitchell, 2014, Overview section, para. 2), are among those that are building news operations. New revenues are flowing into old news operations from entrepreneurs like the *Washington Post's* Jeff Bezos. The use of social media and mobile devices has exploded providing the ability to reach more people with news than ever before. Despite this, most news still comes from newspapers and print ad revenue although its main source of income continues to rapidly decline, down 49 percent from 2003 to 2013 (Mitchell, 2014).

In student media operations, disruptive innovation took longer to be a factor given the largely captive audiences of the typical campus newspaper and its ability to deliver a specific demographic of readers. But sliding revenue for many operations is now a reality and means a contraction of resources and a need to balance shrinking budgets. Cuts to the number of circulation days, online-only publication, and lower salaries are common reactions. Dan Reimold (2014a), author of *College Media Matters*, writes that transitions in student media linked to media disruption follow three rationales: First, that digital and mobile are what student-age readers want; second, employers hire students with new media skills; and third, reducing print production allows student outlets to reprioritize for today's media market in the face of advertising losses and low reserves.

Some examples:

- *The Daily Collegian* at Oklahoma State University shrunk its print schedule from five to three days and changed its format to tabloid from broadsheet in order to focus more strongly on its digital presence. It renamed the print publication *The O'Colly*. "By making this change, we are transitioning from a five-day-a-week newspaper to a 24/7 media company" (Bland, 2014, para. 3), said editor-in-chief Kyle Hinchey.
- *The State Press* at Arizona State University announced in July 2014 it would no longer publish its weekly print newspaper and would be all-digital in the fall. That's after dropping its daily print edition for the weekly in January 2013. In a press release from *The State Press*'s editorial board (2014) adviser Jason Manning said, "We are no longer a so-called 'digital-first' media organization—we are a media organization in the digital age. Our audience and our advertisers are highly mobile and social, and the legacy print product does not serve their needs" ("Press Release," 2014, para. 4).
- The *Columbia Spectator* at Columbia University in New York City goes to a weekly print product and vows to be "more 'daily' than ever" (Abrams, Ouimette, & Lau, 2014, para. 3) as it strengthens its effort to be a 24/7 news source online, freeing "editors and writers to focus on providing our readers with the stories they want through the medium they prefer" (Abrams et al., 2014, para. 4). This change prompted the resignation of John R. MacArthur, President of *Harper's* magazine, from the *Spectator*'s Board of Trustees. In an email, MacArthur said he did not "want to be associated with the disappearance of the *Spectator* into digital oblivion" (Vilensky, 2014, para. 4).

- The *Daily Kent Stater* at Kent State University reduced its print schedule from five to three days per week in August 2014. The move was expected to save almost $38,000 per year in printing and delivery costs and allow the student writers and editors to focus more strongly on producing digital content for *KentWired.com*. The Office of Student Media at Kent State also announced it will search for a Director of Student Media, a new position intended to strengthen strategic planning and new business development.
- The *Mountaineer* at California's Mt. San Antonio College switched its print operations to the online publishing platform Medium and renamed itself *Substance*. Adviser Toni Albertson described the arrangement as "the perfect merge of tech and college journalism" (Albertson, 2014, para. 11) and a way to inspire her "defeated" students, who realized they "were putting out the newspaper for the administration and faculty, and not for the student population" (Albertson, 2014, para. 5) who don't read newspapers anymore. Mountiewire.com, the companion website of the newspaper, will continue to be used for breaking news on campus, sports and general news (Reimold, 2014b).

A LOOK AT THE *OREGON DAILY EMERALD*

One student media group that chose to "embrace the disruption head-on and look for other avenues within the value networks that are ripe for growth and innovation" is the Emerald Media Group (Christensen et al., 2012, Another classic case section, para. 1).

When Ryan Frank began his tenure at the *Oregon Emerald* in February 2011, his title was Publisher of *Oregon Daily Emerald* Publishing Company, Inc. When he left three years later in March 2014, he was President and CEO of Emerald Media Group (personal communication, October 9, 2014). "The idea of having a publisher to me felt so … old. We published stuff, but that was kind of part of the transition of going from a newspaper to a media company" (R. Frank, personal communication, October 9, 2014). Frank said they wanted to think of themselves as a small business, where the person in charge is "the president or CEO so we changed the job title … to make it fit in with that model of who we wanted to be" (R. Frank, personal communication, October 9, 2014).

The student media operation had been through five different leaders in three years including a strike by the student journalists over editorial control. "The board was looking for consistency and direction and vision and they wanted a plan. They were willing for big crazy ideas and risk … the environment was such that people wanted to listen to big audacious plans" (R. Frank, personal communication, October 9, 2014). Frank, an alumnus, describes the University of Oregon as a "mini Berkeley," where the *Oregon Emerald* honors a fiercely independent journalistic tradition. Because of the organizational instability, Frank said the "group of students that we had when I first got there was just covered with scar tissue" (personal communication, October 9, 2014). It took time for the students' hostility toward the business side to cycle out, and for Frank to build relationships based on trust and communication. Once that process took root he concluded "it was not a time to swing for singles and doubles. We needed to swing for home runs. We were swinging for the fences" (personal communication, October 9, 2014).

THE "EVOLUTION": STAGE ONE

Frank wrote what he calls "my first big deal," an impassioned call to action that said, "We have to change like tomorrow. I don't know what it's going to look like. You guys are going to lead the charge. And here's how we're going to do it" (R. Frank, personal communication, October 9, 2014). He and his students hung a sign on the conference room door that said, "If a venture capitalist gave you $750,000 and asked you to build a media company on this campus, what would you build?" Frank constantly asked the students for their insights—they hung paper on the wall under the topics news, advertising, marketing and distribution and scribbled in ideas.

Frank put together a team of student leaders who began taking the broad ideas and built them into a more formal plan; in August 2011 the students presented "Project Evolve" to the board. It was "the first iteration of change" (personal communication, October 9, 2014). The two favorite ideas were an *Emerald*-branded beer and a sex column. Most of the plan was never completed, but it gave them a document to work with and a piece of work they had accomplished. Students saw their "crazy" ideas being adopted and started to get excited.

About halfway through the fall term in 2011, Frank felt students were resisting small changes and decided this "evolve thing is not enough" (personal communication, October 9, 2014). He encouraged the students to forget 92 years of daily publishing history, forget they were a newspaper and challenged them to go back to the venture capitalist's salient question of "what would we build with $750,000?" and answer it in the broadest way possible. Frank said the process "was really about what was best for the audience" (personal communication, October 9, 2014). His top priority was to find the best ways to train students and give them the best opportunities, but the business had to survive financially, as well.

In February 2012, Frank and his top students met with the board. They discussed the state of the newspaper industry, the state of their business. They looked at the status quo and three other options. They looked at financial projections, including an analysis of advertiser frequency. At that point, the *Emerald* had a healthy reserve fund, about 9 months of operating capital, with no long-term debt. The paper had not been profitable for a while, but the year of the change was the best in a decade.

At the end of the presentation, the group told the board "the plane is descending into the lake in the industry. And there's a mountain. And we can guarantee you it's almost certain, if we make no changes, what the outcome is going to be" (R. Frank, personal communication, October 9, 2014). They argued they should "run to the front of the plane and grab the controls" and take the "risk we may crash, we may burn." They would "certainly rather grab the controls and pull out than spin around and let something happen to us" (R. Frank, personal communication, October 9, 2014). In this scenario, the biggest risk is to not change, Frank said (personal communication, October 9, 2014). "When we got done with that presentation, the board looked around and said, 'Great, when can we start?'" (R. Frank, personal communication, October 9, 2014). Thus began "The *Emerald* Revolution."

STUDENT LEARNING

Early on in the revolution, Frank said he would pull students aside and get their feedback individually. In hindsight, he doesn't think that involved them as comprehensively as possible. The students became more engaged when the plan was implemented and he found ways for students to take ownership of things.

Editor-in-Chief Andy Rossback, now a designer with the Marshall Project, said he learned things "that you just can't learn in (journalism) schools today," like "entrepreneurial journalism and lean start up mentality" (personal communication, October 14, 2014). Stories were treated "like we treat developing apps or products. Thinking about how to be agile and how to bring that start up thinking all the way through our organization, from story planning all the way there" (A. Rossback, personal communication, October 14, 2014). Rossback said it quickly became obvious they needed to build an experience around mobile and social—essentially build the whole student media experience around the student, not around the preconceptions of what the newspaper industry at large was doing.

Among the changes they made, all were intended to improve "the revenue and relevance" of the *Emerald*. As part of the revolution, they converted the paper to a tabloid, alternative weekly style newspaper with more of a conversational tone, modeled in part after Portland's Pulitzer Prize-winning *Willamette Week* and created a full-time professional staff position for a programmer, who facilitates student training around technology and helps build web and mobile apps. Frank said the position became a "core part of that team that built the tech, from budgeting to a native iPhone app." It also helped "bring together technology and photo and video in a way that helped the business and diversified revenue and start new companies within the *Emerald* around technology and our other assets" (R. Frank, personal communication, October 9, 2014). The students also started Emerald You, an in-house training program every Friday on leadership, management and journalism; launched a creative ad agency to work with outside clients; started an events planning team to work with campus partners; launched a photo booth business to make money shooting pictures at events; and created an online housing search site.

Frank and Rossback agree journalism schools have difficulty keeping up with the pace of industry change. While many faculty were supporters, Frank said "some of the messaging [students are] getting from their professors is counterproductive to change and it's very indicative of what the whole industry is going through" (personal communication, October 9, 2014). Rossback pointed to a divisive culture between journalism schools and student newspapers. "They each feel like the other is doing journalism a disservice. We certainly had that culture" (personal communication, October 14, 2014). But Rossback believes it's essential to have "the pairing between what you learn in the classroom and the pairing of what you learn on the student paper. You can't have one without the other … there are limitations to what journalism schools can teach students," he said, and "there are limitations to what student newspapers can do to teach students" (personal communication, October 14, 2014).

Rossback said students working for the Emerald Media Group learned how to speak about their ideas, how to get buy-in, how to work, how to be collaborative and how to whiteboard ideas. Twelve months into the "Revolution," he said Frank had "the whole organization thinking more like a start-up" (personal communication, October 14, 2014). At staff meetings, even those students not originally involved in the revolution adopted the new ways of thinking. "There was a sense "this is a real company, this is a real organization, we have a real revenue goal to meet. We have to start acting like that or we're going to evaporate" (personal communication, October 9, 2014).

Rossback originally wanted to be a print newspaper designer, but the environment of innovation and the realities of the job market made him reconsider. "The fact that we operated on this weird agility thing, I was like ok, I'll just change. I am so flexible to switch it up. I don't know if that's something I had ever been exposed to before" (A. Rossback, personal communication, October 14, 2014). "At the *Emerald*, I got to do work that people don't get to do until they're in their 30s or 40s," he said, "strategic decisions about content and business models and stuff that I dream of doing and hope I can do someday … you can't do that in a classroom" (A. Rossback, personal communication, October 14, 2014).

Frank invested the students in a "no nonsense, anything is possible worldview," where if you didn't know how to do something, you were expected to go

Google it and figure it out. "That's really, really valuable" (personal communication, October 14, 2014), Rossback said. "It seems like an old school worldview. Just go figure it out. Just go do it."

Rossback (personal communication, October 14, 2014) said they developed a lot of ideas because they saw a market for it—some ideas were "not as cool," but might enhance the lives of students. The group considered creating a startup incubator for a laundry company that would do laundry for students, leading the *Emerald* students to wonder "Where on earth could you get wrapped up at a college newspaper and end up working on a start up laundry business to help fund the newspaper?" (personal communication, October 14, 2014).

"The *Emerald* set me up perfectly to work here (at the Marshall Project)," Rossback said (personal communication, October 14, 2014). "We're actually a start-up where I may do 100 different tasks in a day that are completely unrelated ... that start-up mentality, that agility, all that stuff. That is what you need to be a journalist today."

THE WESTERN KENTUCKY UNIVERSITY *COLLEGE HEIGHTS HERALD*

At Western Kentucky University, Director of Student Publications Chuck Clark leads a program steeped in tradition, where the newspaper is "still king" in its student media operation, which includes the *Talisman* yearbook. "We are a central part of life at WKU ... more so than at a lot of colleges, we are part of the fabric of what people do and have here" (personal communication, October 8, 2014).

"Media disruption," Clark said, hit WKU at "the same time that 'new director disruption' hit here" with his arrival in June 2012 (personal communication, October 8, 2014). Since then, Clark and his students haven't planned a revolution, but they have made strategic changes to improve overall operations, like revamping the print distribution system, in part to capitalize on the fact that "students are readers by nature."

The challenge in keeping circulation numbers healthy, Clark said, "isn't getting them to read it, it's getting it in their hands. And that's a very different challenge" (personal communication, October 8, 2014). Once the paper is in the students' hands, "they see this is about who I am and what I do and the organizations I'm involved with and what's going on around me and they see value in it" (C. Clark, personal communication, October 8, 2014). But Clark said a static distribution network with boxes and stacks of papers everywhere isn't good enough.

With an already healthy pick-up rate of between 75 percent and 80 percent, Clark organized "street teams" to raise the pick-up rate to a target between 90 percent and 95 percent (personal communication, October 8, 2014). They created

10 distribution hubs in high-traffic pedestrian intersections and pay student organizations to hand out the papers when the most students are out and about on campus. Clark said advertisers depend on the paper's effective distribution and "advertising is the lifeblood of what keeps us in business," since the *Herald* is entirely supported by advertising revenue and gets no direct funding from the University (personal communication, October 8, 2014).

When he arrived, Clark found a "strong, vibrant newspaper newsroom" that has steadily transformed into a "strong vibrant newsroom, not just a newspaper newsroom" (personal communication, October 8, 2014). But, he adds, "There's no question that the newspaper in reality is still king." He credits his students with doing a "wonderful" job of making the newspaper, which is a twice weekly, more enterprising and less reliant on news that's 48 to 84 hours old when it hits print. "They've become a whole lot more effective at using the web and using the apps," Clark said (personal communication, October 8, 2014). And while the app is a good tool for now, with news and a "lot of other cool things on it," he said the nice thing about a college news organization is "we can drop things on a dime. We can try things. And if they work, terrific, and if they don't, so what? It was a good try. There's not a multi-million dollar jobs-at-stake decision there. So we try to be really nimble."

Clark found himself working a bit against his tradition-steeped students, and was surprised, given the newsroom staff changes completely every four years and about 25 percent each year (personal communication, October 8, 2014). "They're still very bound by that *Herald* tradition and in some ways surprisingly reluctant to break away from what their framework says is *Herald* tradition."

He thinks they have become better at making changes—as "they've done more and seen the reaction they've tried new things, tried more things. We are constantly pushing them to try more than they've tried so far" (C. Clark, personal communication, October 8, 2014). Clark's job as an adviser, he said, is to show the students that change is a part of *Herald* tradition, since the *Herald* was one of the first college news organizations to have a website. But then, he said, it "stayed at having a website for 18 years. That's as far as it went."

His students in October 2014 were working on a dramatic upgrade to their website, wkuherald.com, which is built on the TownNews' Blox content management system. They're learning "it's not just as easy as saying 'hey, let's launch a new and improved website'" and are experiencing the frustrations of "it's pegged to roll out in July, it's pegged to roll out in September, it's pegged to roll out in October, now it's pegged to roll out in January. So they very much are getting real world lessons in you just don't' throw something out there and it's going to work."

Overall, he said WKU is doing a lot of the things news organizations should do to combat disruption. He tries to help students who want to pursue journalism jobs understand, in today's disrupted environment, "the more they can do on more platforms, the better chance they're going to have of landing a media job, a news job" (C. Clark, personal communication, October 8, 2014). He also tells students if they can seek out information, put it in context and explain it clearly they're "going to be golden in so many areas. [They're] going to be so well prepared for life after college" (C. Clark, personal communication, October 8, 2014).

Each fall, WKU holds a "big confab" to launch the year. Clark's keynote speech in 2014 was all about "why we're not following the Arizona State lead" to go all-digital (personal communication, October 8, 2014). "Why do we even bother with print?" The answer, Clark said, is "because our audience wants and expects it," and "because it makes a boatload of money." The end of print at WKU would mean the loss of 90 percent of its revenue. "Print in and of itself isn't going to maintain that," Clark said. The print product has to offer readers news that's not "three, four, five days old." It has to "surprise them, inform them, help them see things from a different perspective, make them think" to see the value of print (personal communication, October 8, 2014).

To diversify its business offerings, "College Heights Media," a creative agency was launched and produced welcome guides for the new student union as well as a parent and family weekend guide. The agency charges a creative fee that covers most of its costs, and then sells advertising in the guides. Clark said the efforts are "overwhelmingly successful" (personal communication, October 8, 2014).

Clark is also about to launch a fundraising campaign built around the experiences in student publications that change peoples' lives. Different parts of the campaign will help pay for student internships, direct program support, a capital fund and a leaders fund that will help finance specific student project proposals. Ultimately, Clark said, the journalistic experiences students get of:

> actually practicing it in a real world setting is completely and totally indispensable to whether they're going to be successful, particularly if they want to go into some sort of media. It gets them hired; it puts a roof over their head. Building those kind of muscles and skills—you just can't replace that. (C. Clark, personal communication, October 8, 2014)

CONCLUSION

In *The Innovator's DNA,* researchers found innovative executives (and companies) are different from typical executives, including in the area of "discovery" skills versus "delivery" skills (Dyer, Gregersen, & Christensen, 2011, pp. 33–34). This means the executives with innovative traits are only average at execution of strategy and delivering results, but have strong discovery skills of questioning, observing, networking, and experimenting that help them generate innovative ideas. These ideas help foster new products and new business opportunities that fuel growth. Without these kinds of executives in top management positions, companies become disrupted, stagnant and mature into irrelevance (Dyer et al., 2011, p. 36).

For students who work in disrupted media operations on their campuses, immersion in an environment of discovery helps them prepare to "revolutionize industries and generate wealth" (Dyer et al., 2011, p. 1). At the *Emerald*, Frank said the biggest accomplishment of the revolution was getting students to set big goals for themselves, like pursue their dream jobs. "To me, that was a huge part of it, that they had the self-confidence to do this … they've gone on to do the things they set out to do" (personal communication, October 9, 2014).

For Rossback, the inclusive and innovative environment created at the *Oregon Emerald* during its revolution rubbed off from person to person. "Ryan did it to me, I did it to someone else, the managing editor did it to the news editors." Students invested themselves in "operating in unison around these fundamentals of figure it out, communicate, stand up for your ideas, be involved." It made them care about the enterprise. "And to care," Rossback said, "is really infectious. Really infectious" (personal communication, October 14, 2014).

At Western Kentucky, Clark said responding to disruption means student media operations must "plan for the future—keep your mind open to changing all the time." He feels strongly that part of the "antidote" to media disruption is for student journalists to learn to report in whatever platform is needed and "getting them comfortable in that, insisting that they get to be comfortable with writing, with standing in front of the camera, with building a web package" (personal communication, October 8, 2014).

He sees challenges for his students as they enter the media industry, but he is optimistic about journalism's future. Clark thinks "it can and will survive. To me it's more a matter of figuring out how you do good journalism in a world that changes almost by the hour as opposed to by the year" (personal communication, October 8, 2014).

REFERENCES

Abrams, A., Ouimette, M., & Lau, S. (2014, April 27). *Letter from the editors: Spectator's new direction.* Retrieved November 18, 2014, from http://columbiaspectator.com/opinion/2014/04/27/letter-editors-spectators-new-direction

Albertson, T. (2014, September 14). *The experiment: How one college journalism program said goodbye to all that.* Retrieved October 27, 2014, from https://medium.com/substance/the-experiment-be947b2ba13e

Bland, A. (2014, March 14). OSU's *The Daily O'Collegian will change name, format.* Retrieved November 25, 2014, from http://www.tulsaworld.com/news/education/osu-s-the-daily-o-collegian-will-change-name-format/article_220ba213-d3d8-5267-bc0a-81868b7df9e3.html

Christensen, C. (1997). *The innovator's dilemma: When new technologies cause great firms to fail.* Boston, MA: Harvard Business School Press.

Christensen, C., Skok, D., & Allworth, J. (2012, September 15). *Breaking news.* Retrieved November 22, 2014, from http://niemanreports.org/articles/breaking-news/

Dyer, J., Gregersen, H., & Christensen, C. (2011). *The innovator's DNA: Mastering the five skills of disruptive innovators.* Boston, MA: Harvard Business Press.

Mitchell, A. (2014, March 26). *State of the news media 2014.* Retrieved November 21, 2014, from http://www.journalism.org/2014/03/26/state-of-the-news-media-2014-overview/

Press Release: The State Press goes all-digital. (2014, July 16). Retrieved November 12, 2014, from http://www.statepress.com/2014/07/16/press-release-the-state-press-goes-all-digital/

Reimold, D. (2014a, April 16). "*Dean of the college media business" to student journalists: Stop dropping print!* Retrieved November 28, 2014, from http://www.collegemediamatters.com/2014/04/16/dean-of-the-college-media-business-to-student-journalists-stop-dropping-print

Reimold, D. (2014b, September 16). *College newspaper drops print, first to operate primarily on publishing platform medium.* Retrieved October 27, 2014, from http://www.collegemediamatters.com/2014/09/16/college-newspaper-drops-print-first-to-operate-primarily-on-publishing-platform-medium/

Vilensky, M. (2014, April 29). Trustee quits after Columbia University paper votes to decrease publication. *Metropolis.* Retrieved October 28, 2014, from http://blogs.wsj.com/metropolis/2014/04/29/trustee-quits-after-columbia-university-paper-votes-to-decrease-publication

CHAPTER ELEVEN

"Welcome to The Drew Vault"

Educating Students through College Radio

CHRISTOPHER J. ANDERSON

INTRODUCTION

Since the early twentieth century, colleges and universities throughout the United States have used radio to broadcast a multi-faceted selection of music, campus activities, and educational initiatives. The diversity of college radio programming provides opportunities for university faculty, staff, and students to share their thoughts, personalities, and research over the airwaves or by means of the Internet. These broadcasts highlight the campus experience and provide a medium for students to share their opinions about music, politics, popular culture, and often, whatever crosses their mind. This chapter explores college radio as a broadcast medium for educating the university community and as an outreach tool for certain educational initiatives at Drew University in Madison, New Jersey. The chapter also examines future educational opportunities under consideration as part of the overall radio programming initiative of the Drew Library.

The author's weekly radio show, *The Drew Vault*, is an inexpensive medium used for student education and campus outreach. Each program presents a variety of musical formats, introduces students—as co-hosts and as listeners—to the history of the University, and showcases the activities and collections of the Drew University Library. Drew faculty, staff, students, and alumni participate in the weekly programming and as occasional DJs on the live show. The program is branded as the world's only live radio show dedicated to the study of college history and university archives. The format of the show includes a variety of music from the 1920's to the present and the allotted hour is used to discuss the collections, exhibits, and activities of our library. The show includes segments on the history of Drew University and a variety of guests appear on the broadcast to discuss current events, campus history, and their experiences as researchers with our archival collections.

Mitchell, Seiden, and Taraba (2012) suggest, "Special collections and archives engage and empower undergraduates as well as enhance their learning experiences. These materials are far more than mere survivors of the past; they are truly portals to new ways of learning and thinking" (p. ix). Librarians and archivists are commissioned to protect, preserve, and promote their collections. This is especially true with the faculty and staff of the Department of Special Collections and University Archives at the Drew University Library. Our materials are more than "mere survivors" and our mission or task is to breathe life into and foster the use of our rare books, manuscripts, photographs, and ephemera. As a result, members of our department are proactive in thinking of ways to promote the history of the university as well as our library collections through traditional approaches such as museum exhibits and by means of recent initiatives such as digitization and radio. During our planning, several ideas emerged including initiating an engaging schedule of museum exhibits, hosting a series of meetings with department chairs to showcase our archival materials, as well as creating a new college radio show dedicated to promoting the history of the University and our daily work in the library and archives.

COLLEGE RADIO AS A TOOL FOR EDUCATION AND OUTREACH

The use of college radio by librarians and archivists as a tool for education and outreach is illustrated by the work of campus libraries at The Georgia Institute of Technology, Winston-Salem University, and Washington State University. Limited scholarship is available connecting college and university libraries and archives with radio, in particular the use of college radio as a tool of outreach to promote library special collections and university archives. This chapter joins scholarship investigating these connections.

Ameet Doshi, User Engagement Librarian and Assessment Coordinator for The Georgia Institute of Technology Library, published a chapter examining how librarians at Georgia Tech use college radio to promote the work of the library to faculty, students, and the general public. Doshi (2012) confirms the use of college radio shapes user perceptions of the library since on-air "conversations about the library are happening live and broadcasting over the airwaves gives listeners a sense of authenticity about the breadth of serendipitous encounters between library users and library staff" (p. 25). The library produces a live show on Atlanta-based campus radio station WREK 91.1 FM. *Lost in the Stacks* is hosted by librarians Doshi and Charlie Bennett. The program first aired on January 24, 2010 and is marketed as "the One and Only Research Library Rock 'n Roll Show!" ("Lost in the Stacks," 2014). The goal of the show is to "connect the Library with Georgia Tech students, staff and faculty on WREK." The format includes a variety of music including Indie Pop, New Wave, and Rock particularly selected for the theme of the day's program. The hosts also interview the people of Georgia Tech as well as outside guests who discuss news and topics related to libraries. The program broadcasts every Friday at noon EST over the air to the Atlanta area and is also streamed live on the Internet. The show is made into a Podcast and is also promoted through both Facebook and Twitter. Doshi confirms listener evaluations, "suggest that the radio program has been successful, low-cost, high impact collaboration with student media that helps shape user perception of the Georgia Tech library and library staff in new and positive ways" (p. 26).

In 2006, the C. G. O'Kelly Library at Winston-Salem University in North Carolina launched a call-in library radio show titled *Let's Talk Learning Spaces*. The show is hosted by Forrest Foster, Access Services Librarian and Information Commons Coordinator, and is available as a Podcast on BlogTalk Radio. Foster uses the program to discuss news of the library including concerns or ideas regarding the uses of learning spaces in university settings. He confirms his radio show provides a media platform for "collaborating and sharing information about best practices for policies, services, research, and technology in regards to learning spaces, i.e. learning/information commons" ("Let's Talk Learning Spaces Show," n.d.). He also advertises his show on his Twitter account, @forrestfoster.

On the campus of Washington State University Reference Librarian Patrick Wohlmut hosts the library radio show *Voices Carry*. The show streams live on the Internet every other Thursday from the radio studio of KOUG. The talk-show format includes conversations with guests on a variety of topics including protecting your online content and issues more particular to libraries including how to use materials located in the library to help select topics for research ("Library Radio Show Airs 10/3/13," n.d.).

While the radio broadcasts of Doshi, Foster, and Wohlmut examine the work and activities of the campus library, few university libraries appear to use college radio as a medium to promote their archival and special collections materials. After an extensive search for evidence in print, and on the Internet, few examples emerged of librarians using this medium to promote their work and collections. These programs were done more as public service announcements promoting particular libraries and archives than as live on-air shows in real time. Though less has been written on the use of radio to promote college history and archives several examples demonstrate how the medium can be used effectively.

Mumford (1970) notes how archivist Charles Braibant used radio broadcasts in 1949 to bring attention to the underutilized and underfunded National Archive of France. During the 1960s the Government Archivist of Jamaica used the radio to discuss the state of archives in the country. A transcript of the addresses is available in the publication *Our Archives: Six Broadcast Talks by the Government Archivist* (Black, 1962). In 1964, Mary Givens Bryan, Archivist for the State of Georgia, gave a paper demonstrating how she had used a variety of approaches, including radio, to promote the state collection at the Georgia Archives Building (Bryan, 1964). Pederson (1978) notes a survey by the Committee on the Wider Use of Archives of the Society of American Archivists was conducted in 1976 to discover ways archivists promoted their collections to the general public through a variety of means of outreach. Several libraries used forms of mass media including television and radio to promote their collections. These promotional tools were brief, previously recorded public service announcements describing archival programs and services (Pederson, 1978).

A final example is a radio production by the Bird Library at Syracuse University. In 2011, the library produced 90-second radio spots called *Sound Beat* comprised of historical recordings from the Belfer Audio archive. The spot aired daily at 3:30 p.m. on local public radio and included commentary that contextualized each particular recording (McLaughlin, 2011). Librarians also created a Facebook page to promote their recorded collection ("What is Sound Beat?" n.d.).

THE DREW VAULT RADIO SHOW

In "Embracing Web 2.0: Archives and the Newest Generation of Web Applications," Mary Samouelian (2009) suggests librarians and archivists consider radio or "radio-type shows" such as Podcasting to bring attention to archival collections. This approach to connecting students with library and archival materials was the motivation behind the start of my new radio show dedicated to the study of university history and the archival materials at the Drew University Library. This section examines the origins of *The Drew Vault* radio program and considers how the show functions as a tool for education and outreach to the students of Drew University.

In 2013, the Department of Special Collections and University Archives was formed during restructuring of the Drew University Library. As Head of the Department, I was charged to think creatively regarding how to promote our work and collections both on- and off-campus. Several possible organizational venues were available to promote the work of the library including the various student clubs, the *Acorn* student newspaper, and the college radio station WMNJ. I decided to investigate how the library might use the college radio station as a broadcast medium to promote the history and archival collections of the university.

Drew University radio station WMNJ "The Forest" was chartered in 1961, as an AM-band station with a signal strength reaching only the buildings within the campus. The first broadcast to the campus community took place on April 4, 1961, when the President of Drew made several remarks on-air concerning the founding of the station. By the early 1980s the station had moved to FM and the signal reached beyond the campus and into the local Madison area. By 2012, the station had moved online and currently streams both live programming and automated music twenty-four hours a day, seven days-a-week at wmnjtheforest.com (E. Buchholz, personal communication, December 30, 2013). In 2013, the new student union, The Ehinger Center, opened as a gathering space for the student population. The facility was largely funded by a Drew University alumnus who had been a DJ during his undergraduate years at the university. As a result, the studio booth was prominently placed near the entrance of the student center, complete with a large viewing window allowing visitors to look in on DJs as they broadcast from the booth.

During the summer of 2013, I decided to plan and host a radio show dedicated to student learning and outreach for the library and university archives in time for the fall semester. I submitted a DJ application to the executive board of the radio station which then approved my show for the fall semester. The show was originally titled *The 80's Vault* with a timeslot of Wednesday evenings at 7 p.m. Each broadcast was used as a platform to play a variety of music from the 1980's, educate students on the history of Drew University, and discuss the current work of our special collections and university archives department. In January 2014, *The 80's Vault* was rebranded as *The Drew Vault* (@TheDrewVault) and continues the same format though allowing for more musical diversity. The program is now marketed as the world's only live radio show dedicated to the study of college history and university archives and the music selection was broadened to include songs from the 1920's until the present. This allows for more flexibility with the selection of music and also presents opportunities to connect with listeners who have a greater variety of musical interests.

Educating Students through *The Drew Vault*

Ana Friz (2008) notes in *Cinema Journal*, "Campus/community radio provokes the audience into listening for the unexpected while encouraging them to come to the station and become programmers, thus beginning a kind of circle of transception between the listening community and the radio station" (p. 140). *The Drew Vault* radio show echoes this approach by educating students through on-air discussions of the history of the university. The show also brings undergraduate and graduate students into the studio as experiential participants as DJs and as programming assistants. Students are invited to participate on live broadcasts as guests and they assist, from time-to-time, with creating content for the show by suggesting topics for discussion. When students appear on the show they'll read portions of the script for the broadcast or I will ask them questions concerning how they envision former students experienced campus life and activities during the early years of the campus. These broadcasts provide opportunities for students to learn about the content of the show's programming and also provide practical experience by learning how to host a show, create content such as scripts and musical selections, and to run the sound board and computer system that generates the broadcast.

Educating students through programming and participation is demonstrated from the following commentary by Drew students who have participated in the show. Each has been a listener to the program and a guest DJ. Brian Shetler, a Ph.D. candidate in History and Culture at Drew, recently remarked,

> *The Drew Vault* is a valuable outreach tool for archives and special collections at Drew University. Through an aural interaction with the materials, the show is able to connect listeners to primary sources that bring the history of the university to life. Beyond highlighting interesting and significant individual items, the show also brings attention to the archival and special collections holdings that can be found at Drew. Perhaps the best effect of the show is that it is able to reach an audience (namely students) that might otherwise not know about these materials and the value that they hold. (B. Shetler, personal communication, December 17, 2014)

Undergraduate students also participate in the show as *The Drew Vault* provides both an avenue to educating the student as well as opportunities to learn how to broadcast their own shows. Drew sophomore student T. J. Chang, who now has his own Drew radio program, recently confirmed,

> Learning more about the history of Drew gives the undergraduate experience a whole new outlook. Knowing the history makes one appreciate the hard work and dedication the founder and benefactors put into creating what Drew is today. After all, without them Drew would not be here today and I would not have met the wonderful people who roam Drew. (T. J. Chang, personal communication, December 7, 2014)

The program also invites off-campus researchers into the studio who are visiting the Drew archive for their own scholarship. By appearing on the program, researchers are provided with a platform to discuss their work and to promote the archival collections to the listening audience. Inviting researchers to share their experiences at the Drew archive on-air educates the campus community on our collections as many students are likely unaware of the rich resources available at the archive.

Over the past two years I have invited researchers using our archival materials to join me at the radio studio to co-host a live show. During the broadcast researchers share details about their work and confirm how the resources of the Drew archive have been essential for their scholarship. A recent guest is an Assistant Professor of History at Rust College in Holly Springs, Mississippi. Dr. Marco Robinson discussed his research that includes a project on the history of Rust College. He enjoyed his experiences at the archive and participating in the radio broadcast, later noting,

> As a historian and college professor, I always promote and utilize innovative methods to engage my students, colleagues and the general public with the discipline of history and conducting research in the field. The public perception that history is dry, date-driven and irrelevant is a product of college students' and the extended community's lack of exposure to history repositories and archives. The live radio show I participated in at Drew University gave me an opportunity to address both issues. Thus, *The Drew Vault* radio show, from a pedagogical and community service perspective, is on the cutting edge of engaging the community with archival materials while highlighting the history of Drew University in Madison, New Jersey. (M. Robinson, personal communication, December 8, 2014)

A second outside researcher who appeared on the show is a Ph.D. candidate at the College of William and Mary in Williamsburg, Virginia. Christopher Jones discussed his research and scholarship and noted he had spent considerable time at the Drew archives facility completing an extensive investigation into the 19th century intersections of Methodism and Mormonism in the United States. His research at the Drew archive led to the publication in 2012 of a scholarly article in *BYU Studies Quarterly* journal that drew the attention of several high-level leaders of the Church of Jesus Christ of Latter-day Saints (Jones, 2012). The article demonstrated a small, yet significant historical change to two sections of *The Doctrine and Covenants*, one of the authorized scriptures of the Mormon tradition. His research altered the original text from 1839 and corrected the historical context in the most recent edition of the volume.

During the show I asked him to discuss his findings from our archive and to share how his research at Drew had led to a historical footnote in Mormon scripture. *The Drew Vault* radio show provided Christopher with an opportunity to discuss his doctoral research while at the same time the broadcast helped educate our student listeners and promoted our archives collections. Christopher responded,

> Being a guest on Chris Anderson's radio show, *The Drew Vault*, was a great opportunity; scholars dream of sharing their research with a broader audience, and this provided a rare chance for a young scholar like me to do so. That the audience was primarily members of the community in and around Drew University, where I've visited over a dozen times in the last decade for archival research trips, was especially meaningful to me, and made me feel like a real part of that community. The article I published in 2012 that was based primarily on archival and print sources housed at Drew's Methodist Archives & History Center has attracted more attention than any other of my scholarly efforts to date, and much of that

> exposure is due to Chris's kind invitation to have me as a guest on his show. (C. C. Jones, personal communication, December 8, 2014)

The show was well-received and attracted listeners from on-campus as well as from North Carolina, Ohio, and Texas.

FUTURE OPPORTUNITIES FOR EDUCATING STUDENTS AND ALUMNI

Steven McClung (2001) notes the importance of connecting alumni through college radio, "Among alumni … staying in touch with the university is also the main reason for using the college radio website" as the website and campus radio broadcasts "strengthen ties to the university where they spent time, and quite possibly to which they have a strong allegiance" (pp. 69–70). This link is a significant reason why we have created programming that connects current students and alumni with the history of the university. Our intent is to also spotlight particular archival items that current students may appreciate and historical campus events that alumni may reflect back upon during their time on campus. Two specific components of using radio to educate students and alumni in the future include the *George M. Eberhardt Historical Audio Series* and the *People of Drew* series, one launched in early 2014 and the other is currently being planned as a future program.

George M. Eberhardt was a sound engineer at Drew University for over thirty-five years, retiring at the age of ninety-nine. During his time at Drew, Eberhardt recorded hundreds of on-campus activities that included well-known speakers, faculty lectures, and student events ("Drew Remembers George Eberhardt," n.d.). We decided to honor the legacy of Eberhardt while promoting his many original recordings housed in our university archive. The *George M. Eberhardt Historical Audio Series* commemorates his work at Drew. In February 2014, we planned and hosted a broadcast of *The Drew Vault* commemorating the 1964 campus visit of Dr. Martin Luther King, Jr. The show included a biographical segment on King, details of his day on the Drew campus as recorded in the campus archival collection, and the original recording of Dr. King's address ("Martin Luther King to Be Heard Again," n.d.). The broadcast aired fifty years to the month of King's original speech and was played online and over the public address system of the new student center. I contacted the Department of Alumni and Parent Relations and asked them to send an email to all Drew alumni encouraging them to listen to the broadcast. During the show we made a request for alumni to donate historical items such as photographs, recordings, and ephemera related to King's visit. The broadcast not only promoted our archival recording of King it also helped build the university archive collection related to the event. The broadcast also invited alumni to think back on their days at Drew and provides opportunities for the Drew Library to collaborate with the Department of Alumni and Parent Relations.

More recently, a second broadcast from the Eberhardt collection commemorated the 1962 visit to Drew of Indira Gandhi, former Prime Minister and social justice advocate from India ("Drew Library Radio Show to Broadcast Original," n.d.). Like the King broadcast, the show examined Gandhi's day at the Drew campus through the archival record and we played the original speech for listeners. For this broadcast we collaborated with Dr. Karen Pechilis, a member of the faculty of the Drew University Department of Comparative Religion. Dr. Pechilis joined us in the radio booth to provide a biographical sketch of the Indian reformer as well as place Gandhi in the political and religious context of India during the 1960s. The broadcast was aired over the Internet as well as over the public address system of the student center.

CONCLUSION

College radio provides interactive windows of opportunity to educate students, alumni, and the larger university campus community. As one of many campus student organizations that function as crucibles of learning, college radio can be used as a teaching tool. *The Drew Vault* radio show is one example of a way to implement a college organization to inform students and alumni on the history of a university while also bringing attention to the work and collections of a college library and archive. Each program provides students with avenues to create and interact with the content of broadcasts while also fostering first-hand opportunities to hone their skills in public speaking and using studio technology.

REFERENCES

Black, C. (1962). *Our archives: Six broadcast talks by the government archivist.* Kingston, Jamaica: Government Printer.

Bryan, M. G. (1964). The Georgia archives building—A case study in promotion. *American Archivist, 27*(4), 499–501.

Doshi, A. (2012). Lost in the Stacks: The Research library rock 'n' roll radio show! In L. Snavely (Ed.), *Student engagement and the academic library* (pp. 23–30). Santa Barbara, CA: Libraries Unlimited.

Drew Library Radio Show to Broadcast Original 1962 Recording of Indira Gandhi. Retrieved December 30, 2014, from http://www.drew.edu/library/2014/11/vault-i-gandhi

Drew Remembers George Eberhardt, Former Sound Engineer and Campus Fixture. Retrieved December 30, 2014, from http://www.drew.edu/news/2014/05/05/drew-remembers-george-eberhardt-former-sound-engineer-and-campus-fixture

Friz, A. (2008). Re-Enchanting radio. *Cinema Journal, 48*(1), 138–145.

Jones, C. C. (2012). Mormonism in the Methodist marketplace: James Covel and the historical background of *Doctrines and Covenants* 39–40. *BYU Studies Quarterly*, *51*(1), 67–98.

Let's Talk Learning Spaces Show. Retrieved December 30, 2014, from http://www.blogtalkradio.com/okellylibraryradio

Library Radio Show Airs 10/3/13. Retrieved December 30, 2014, from http://library.vancouver.wsu.edu/news/library-radio-show-airs-10313

Lost in the Stacks. Retrieved December 30, 2014, from http://www.wrek.org/lostinthestacks/

Martin Luther King to be heard again at Drew, Feb. 26. Retrieved December 30, 2014, from https://www.drew.edu/library/2014/02/martin-luther-king-to-be-heard-again-at-drew-feb-26

McClung, S. R. (2001). College radio station websites: Perceptions of value and use. *Journalism and Mass Communication Education*, *56*(1), 62–73.

McLaughlin, P. W. (2011). *Syracuse University Library Debuts 'Sound Beat' Radio Show, highlights Belfer Recordings.* Retrieved December 30, 2014, from http://syr.edu/news/articles/2011/sound-beat-03-11.html

Mitchell, E., Seiden, P., & Taraba, S. (Eds.). (2012). *Past or portal? Enhancing undergraduate learning through special collections and archives.* Chicago, IL: Association of College and Research Libraries, a Division of the American Library Association.

Mumford, L. Q. (1970). Archivists and librarians: Time for a new look. *American Archivist*, *33*(3), 269–274.

Pederson, A. E. (1978). Archival outreach: SAA's 1976 survey. *American Archivist*, *41*(2), 155–162.

Samouelian, M. (2009). Embracing web 2.0: Archives and the newest generation of web applications. *American Archivist*, *72*(1), 42–71.

What is Sound Beat? Retrieved December 30, 2014, from http://soundbeat.org/about/

CHAPTER TWELVE

The Shock OF THE New

A Case Study of Technological Innovation at a Student-Run University Television Station

K. MEGAN HOPPER AND JOHN HUXFORD

INTRODUCTION

In 2014, the School of Communication at Illinois State University (ISU) opened the Social Media Analytics Command Center (SMACC), a state-of-the-art computer lab that can be used to follow and analyze public conversations from social media outlets—including Facebook, Twitter, Instagram, Reddit, YouTube, and Google Plus—as well as monitor popular discussion forums and blogs. The new technology quickly caused ripples across both the school's Journalism program and at the campus student television station (TV-10), which is part of that program.

Within months, SMACC training was being integrated into classroom discussion, while TV-10 launched a new 30-minute documentary show investigating topical social issues in a way that relied heavily on, and was presented from, the SMACC laboratory. Our study explores the full impact and ramifications this new technology is having on the teaching and training of Journalism students at ISU and on the production of TV-10; an impact made all the more acute by the fact that the school's Journalism program is currently moving to a new convergence curriculum.

Through a case study involving critical analysis, direct observation, and interviews with students and faculty advisors currently engaged with the SMACC, we seek to examine how one student-run television station is incorporating cutting edge technology to implement convergence practices. Our hope is that the lessons offered by this analysis will provide insight to other journalism educators who are dealing with the exciting opportunities—and headaches—of preparing their students for a rapidly converging media environment.

IMPACT OF SOCIAL MEDIA ON THE JOURNALISM PROFESSION

Since the arrival of the Internet, the journalism profession has been faced with the increasing need to evolve. Convergence has become the buzzword; however, it has largely been defined poorly, with varied conceptualizations (Kolodzy, Grant, DeMars, & Wilkinson, 2014). A central impact of convergence has been its ability to allow the consumer to contribute to news content through the use of social media (e.g., Facebook, Twitter, Instagram), which has prompted the need for the journalism profession to re-conceptualize the role of its audience. Through social media, content consumers are transformed into content producers (Kraeplin & Batsell, 2011; Stassen, 2010), which may then prompt the need for journalists to become responsive to such content, as well as to serve as a form of moderator, or in some cases, a mediator of that content. As such, Dimitrov (2014) argues social media has transformed the journalist from performing the role of watchdog to that of "guide dog" (p. 7).

Beyond this, scholars and professionals have long argued that journalists needed to establish a presence in the social media conversation, both contributing a voice as well as listening in on those contributing (Overholser, 2009; Stassen, 2010). New technology is affording journalists the opportunity to tap into social media in order to not only generate leads for stories and to find relevant sources (Picard, 2009; Skoler, 2009; Stassen, 2010) but also to grow, and connect to, their audiences (Burzynski Bullard, 2013). "Through constant interaction with users, journalists develop a different type of relationship than the 'arms-length' connection created by traditional mass communication" (Stassen, 2010, p. 6).

However, there have also been concerns over some of the potential disadvantages of a reliance on social media as a journalistic tool. For example, Kolodzy et al. (2014) point out that although the digital tools, such as social media, that are used to deliver the news may change, the essential purpose of journalism will not. "The touchstones of journalism and more importantly the public's expectation of it involve narrative, information, verification, human interaction, connection, and relevance" (Kolodzy et al., 2014, p. 202). Thus, a focus on utilizing social media over, or in place of, such touchstones may result in a form of journalism that is not of the most benefit to the consumer. Furthermore, Stassen (2010) argues there is a larger global issue of a digital divide that is important to take into account, as social media in some societies "is reserved for a privileged minority who has access to the Internet, and who is somewhat computer-savvy" (p. 8). If journalists begin to rely on social media as a pulse of the community in generating story ideas and finding source material—as well as the central means to interact with the community they are serving—many voices may go unheard.

INCORPORATING SOCIAL MEDIA IN JOURNALISM EDUCATION

Along with the changes the journalism profession has had to make in order to adapt to the digital age, journalism educators have also been faced with making alterations to their curriculum in order to best prepare students for a career in news. Many journalism programs across the globe have thus sought the most effective ways to update and implement their program's curriculum in order to address convergence (Castaneda, Murphy, & Hether, 2005; Kraeplin & Criado, 2005). These changes are certainly needed; as Kraeplin and Criado found, both television and newspaper executives in 210 of the largest U.S. media markets desire new hires to have the skills to write stories for multiple platforms. However, finding the right balance between providing instruction on convergence skills and teaching traditional journalism skills has been a work in progress. More recently, the need to incorporate social media training into convergence journalism curriculum has also emerged.

Since the birth of social media and its increasing popularity among adolescents and young adults, scholars and educators alike have realized the need to examine the role of social media in education (see Vanwynsberghe & Verdegem, 2013). Certainly university students in all disciplines may benefit from lessons in social media literacy, both in their chosen professions and in their more basic role as citizens of the world. However, for the journalism student, education on how best to leverage social media as a journalistic tool appears to be a crucial component of their professional success after graduation.

While most in academia, as well as in the industry, concur there is a need to integrate social media instruction into journalism curriculum, Bor (2014) argues that very little research has been conducted on how it has been incorporated thus far, or how effective such instruction has proved to be. Consequently, the best approach to incorporating social media instruction in ways that are most beneficial to the journalism student is still largely unknown. Our case study of the student-run television station on the campus of ISU seeks to address this need.

UTILIZING SOCIAL MEDIA ANALYTICS AT ILLINOIS STATE UNIVERSITY

The School of Communication at ISU recently opened the Social Media Analytics Command Center (SMACC) to integrate social media into the learning experience and equip students with the most up-to-date knowledge on how to leverage social media data analysis in their chosen career paths. The technology powering the SMACC, Nuvi, fosters social media listening, engagement, analysis and reporting, and offers students at ISU the opportunity to engage with a technology that very few universities currently possess. The project cost approximately $72,000 in total, which funded the renovations to make space for the center, all of the necessary equipment, and a subscription for the Nuvi software.

Specifically, the SMACC allows for individuals to follow keywords and topics to monitor on multiple touchscreens exactly how, and how much, these terms are being discussed on a variety of social media platforms. The technology also enables users to analyze the reach and spread of a particular media posting to see how many others are actually viewing the post and how frequently that post is being shared by other users, to detail how positive or negative that post may be, as determined by keywords, and to pull up specific posts from a cluster. According to Dr. Stephen Hunt, professor and executive director of the School of Communication at ISU, the initial idea to construct the SMACC came in response to the need for the school to address both the rapid rise of media convergence and the growing impact of social media. "The idea for the SMACC, from the beginning, was to build a space that would allow us to train all of our majors for new media competency," Hunt explained. "We also viewed the SMACC as a space to reflect critically on social media and teach students about the dark side of social."

The SMACC provides many opportunities for convergence both within and outside of the School of Communication at ISU, allowing the school to engage in a variety of interdisciplinary partnerships with the university's Technology and Psychology departments, the College of Education, and others. It has also positioned the school as a national leader in social analytics, inspiring other universities and corporations to investigate how they can leverage social media analysis and/or potentially create a social media command center of their own. According to Hunt, the School of Communication has consulted: with Miami University Middletown campus; Kutztown University; and Ohio University's Scripps College of Communication, to name but a few academic institutions, and has developed partnerships with State Farm Insurance, Pinnacle Actuaries, Country Financial, Caterpillar, as well as other corporations.

Although the SMACC is available to students in all disciplines on the campus of ISU, it is particularly relevant for those pursuing a degree in journalism. Specifically, social media analytics is useful to journalists in providing them with a sense of public sentiment and in identifying what people care about in regard to a given issue; i.e., what the public's main concerns are. Furthermore, the technology housed in the SMACC provides journalists with an additional resource to find potential story topics, as well as sources relevant to any given story. It allows the user to map geographically where sources are located and how much they are impacted by the specifics of the story topic. As such, the journalism program in the School of Communication has sought to explore ways in which the SMACC could be incorporated into the program's recently revised converged curriculum, a revision that broadened the range and depth of training to include a wider skill-set across media, as well as adding new classes in journalism theory. One key use of the SMACC, and the focus of the present study, was incorporation of the technology into the training and production of ISU's student-run television station, TV-10, which is housed within the journalism program course offerings.

TV-10

TV-10 is the only fully-staffed TV newsroom in Bloomington/Normal and doubles as a working laboratory and an academic unit within the School of Communication at ISU, providing students with hands-on training in broadcast journalism and production. Students taking journalism classes fill every role at the station: reporters, producers, anchors, photographers, and editors, writing the stories, creating the graphics and producing the weather forecast, all under the guidance of instructors with 25 years of news experience. Eight live shows, more than any other university in the state, are aired each week—five daily news and sports broadcasts and specialized shows covering entertainment, sports and, most recently, topical issues. The station is a regular recipient of awards from the Illinois News Broadcasters Association and has seen its graduates fill positions across the state and beyond.

The implementation and integration of the SMACC at ISU prompted several students and the faculty advisors at TV-10 to create a new program that would incorporate the SMACC technology and be broadcast from inside the center. *The Bottom Line* was conceived and first broadcast via TV-10 in the fall of 2014, with students given substantial autonomy in planning and producing the show.

In the following sections, we will describe how TV-10 incorporated the SMACC into the production of this new, flagship program, and its potential impact on student learning. Specifically, our analysis has sought to answer the following research questions: How was this technology affecting the way news is produced and conceived, both in the classroom and at the student TV station? What exactly are the benefits of adding social media dissemination and analysis to the learning experience? What skills are provided to students who are working with such technology, and how will these benefit them on the job market and in the professional world? Equally important, what are the potential problems and disadvantages that may come with such a focus?

METHODOLOGY

The present qualitative inquiry employed an instrumental case study to examine the particular operation of TV-10. According to Stake (2000), such a study involves looking into a particular case, such as the operation of TV-10 and its use of the SMACC technology, to "provide insight into an issue or to redraw a generalization" (p. 437). Data may be collected in a multitude of ways, such as in the form of in-depth interviews and direct observation, which we gathered as evidence for the present study. Observations may be directed at events/activities that can provide relevant information. "For instance, if the case study is about a new technology, observations of the technology in the field setting can be highly valuable" (Baskarada, 2014, p. 14).

Our semi-structured one-on-one interviews were conducted with the students and faculty directly involved with the production of *The Bottom Line* and with key faculty responsible for bringing the SMACC technology to campus and maintaining its daily functioning. In addition to the interviews, we collected evidence for our case study through direct observation of the production of *The Bottom Line* and the use of the SMACC's resources for the program.

Prior to the analysis, the content of each interview captured via audiotape were transcribed and typed into a word processing file. The data were then interpreted based on a thematic analysis (Owen, 1984). Themes were identified when their meanings recurred, repeated, and were discussed forcefully across the data.

FINDINGS

In our interviews, it was clear that those students who were engaging with SMACC technology were pleased to have been given the opportunity to do so, a sentiment perhaps best summed up by one student: "I love it—it's a lot of work but it's worth it" (Anonymous student 2, personal correspondence, April 2015). Similarly all the students creating *The Bottom Line* spoke with enthusiasm about the process:

> It's always been difficult to rope [social media] in and get information from—until now. Now we can actually analyze what people are tweeting, how many people, and how much of an impact it actually has. (Anonymous student 1, personal correspondence, April 2015)

> We look at hashtags, word clouds, different words that people use—like the most common words that you can look at—what percentage of comments are positive or negative on a subject … it's a really cool program and you can do so much with it. (Anonymous student 3, personal correspondence, April 2015)

> A lot of this you could take into just about any job in the communications industry. This is definitely something that a lot of students need to take—learning about the SMACC. (Anonymous student 4, personal correspondence, April 2015)

Furthermore, a recent graduate, and one of the first TV-10 students to work in the SMACC, felt he had benefitted tremendously from his work on *The Bottom Line*. "I got to try things most people don't get to, and I walked away with skills most people in my shoes don't have" (Anonymous recent graduate, personal correspondence, April 2015).

Job Prospects

Overall, the students were positive about the effects that the SMACC technology would have on the news industry:

> I think (this technology) is good for journalism because it lets us see what consumers are saying, what people want to hear, what their opinions are, what they care about, what we should cover as people care about certain issues. (Anonymous student 3, personal correspondence, April 2015)

However, unsurprisingly perhaps, uppermost in the minds of most of our interviewees was the way that learning these technological skills might increase their value on the job market. For example:

> I think the big thing is how it will help my career … getting that first job. Companies are looking for younger people who can use social media and have experience in programs like these. When you're saying that you can use these programs that analyze the data—it's a huge resume booster, I think. (Anonymous student 3, personal correspondence, April 2015)

The technology was seen as being versatile enough for this to hold true even when the student didn't directly envisage a career in journalism. Another student, while aiming at a position producing entertainment media, was equally positive:

> Social media is very up and coming. Even though it's been out there for a while, it's always changing. Even if I'm not on camera using this, I'll still definitely need to look at this kind of software in whatever company I work in to find out what people are talking about, what are their likes and dislikes—you have to cater to an audience or a customer if you want to make money. (Anonymous student 4, personal correspondence, April 2015)

One student who was currently on the job market, felt positive about her prospects, given her experience on *The Bottom Line* and working with the SMACC technology.

> As stations are solidifying their social media presence, I feel confident in the skills I learned and that we were innovators with what we did with the SMACC. I feel that I can then help stations implement new ideas and directions to further establish their social media presence. (Anonymous student 5, personal correspondence, April, 2015)

In contemplating their future prospects, most of the students were quick to reference the case of a former *Bottom Line* producer who had gone directly into a top TV news station after graduating. He is now a reporter/videographer with an emphasis on using social media and believes his experience working in the SMACC set him apart from the other candidates for the position. "I was walking out of college with something I could use to sell myself … I was the one and the only new-broadcast-major-college-grad ANYWHERE with this experience. That felt incredible" (Anonymous recent graduate, personal correspondence, April 2015).

Interestingly, several of the students saw the SMACC technology, and the skills they were learning, as an empowerment towards what might be described as enterprise journalism:

> I'm going to freelance report when I graduate and sell my stories and potentially travel so this is the time to prepare. (Anonymous student 1, personal correspondence, April, 2015)

> This is the big opportunity that journalists have got right now. If you harness social media, you don't even need to be working for a news station or a corporation or a production office right now. If you harness it and promote yourself right through social media, you can be an entrepreneur of sorts and build your own web site and your work, profile and personality can get out there. (Anonymous student 2, personal correspondence, April 2015)

It is an ambition that faculty member and TV-10 News Director Laura Trendle Polus regards as increasingly realistic: "We have seen some of our most recent crop of graduates go into some of these new ventures—things like web only journalism," she says. "The ability to reach audiences yourself—without a printing press or a transmitter—is now out there. Sites are being developed right now where they will go to work. Or they themselves may develop where they go to work" (Personal communication, April 2015).

Privacy Issues

The fact that most of our interviewees had an eye towards their future employment is a likely explanation for why issues surrounding SMACC technology and privacy were high on their list of the system's potential disadvantages.

While the possibility of the technology increasing the tendency towards "gotcha" journalism was raised, the impact that unearthing incautious tweets might have on personal reputation and job prospects resounded more strongly. The following comments illustrate the feelings of many of our interviewees:

> Before this, I never realized what could be picked up online. I have now privated my accounts so you can't see any of my tweets in the SMACC. But the second that's unprivated, they can see your picture, pull up all your information; if you have your GPS locator on, they can see where exactly they were tweeted from. (Anonymous student 2, personal correspondence, April 2015)

> Everything you do can be so heavily monitored that they can go back and see what I tweeted about two years ago when I was a stupid 18–19 year-old. For a future job, that could really impact your future. It does make you so much more aware of how much is out there. It kind of keeps me up at night. (Anonymous student 4, personal correspondence, April 2015)

Traditional vs. Convergence Journalism Skills

It is perhaps inevitable that, given the context of these interviews, the students we talked to were more focused on the issues and abilities accompanying new technology, rather than on the potential, often negative, impact such technology might have on the teaching of traditional journalism skills (see Castaneda et al., 2005; Kolodzy et al., 2014).

The move to a convergence journalism curriculum at ISU has brought ever more technology into the classroom. This, arguably, has already seen unexpected repercussions. One faculty member told of his surprise when, in a snap challenge he issued to his class to go and find a news story, the students—rather than rushing from the classroom as in previous semesters—instead wanted to stay in their seats and surf the campus web sites for story ideas. As this teacher pointed out:

> Mining web sites like this is fine. But I think it's undeniable that you get a much richer range of stories by *not* only relying on the web, which invariably just means rehashing stuff that is already out there. Finding new stories—big and small—is usually about human interaction. (Anonymous professor, personal correspondence, April 2015)

One can argue that the SMACC technology might lead to the same sort of techno-dependence, with student journalists taking the easy course of snatching quotes from social media—with all its limitations in terms of variety of sources (Stassen, 2010)—rather than search for less convenient voices. Indeed, such techno-dependence seems evident in the following quote from one of our student interviewees: "I wonder a lot of times how people got around not having social media to report. Without it would be like starting from scratch and learning how to report all over again—especially when finding sources" (Anonymous student 5, personal correspondence, April 2015).

The recent graduate we spoke to who is now employed at a large television station spoke of his misgivings about his current work with social media technology:

> It makes reporters lazy. I don't have to go talk to anyone or even leave the station. I'll just set up my tripod in front of my desk and make a few clicks. This is a challenge I'm still trying to overcome. (Anonymous recent graduate, personal correspondence, April 2015)

Further, while being an advocate for journalism's employment of social media, he also saw other limitations to its use, with the platform generating material that lacked "the human element": "You can never get the same emotional reactions from people online. They don't yell, they just type capital letters. They don't cry, they just use emoticons. The internet is rich in everything except emotion" (Anonymous recent graduate, personal correspondence, April 2015).

Yet, arguably, these dangers have been largely sidestepped at *The Bottom Line*. As Trendle Polus explains:

> It has been a conscious decision not to simply rely on social media … The idea of *The Bottom Line* is we take a topic each week and examine it through many voices and many lenses, social media being just one of those. Others are campus voices, the expert voice and the local voice versus the national or international voice—so that's been a useful prism. (Personal correspondence, April 2015)

More broadly, there is a clear tension between maintaining the standard of teaching journalism's foundational skills—including grammar, visual composition, interviewing techniques, and accuracy—and the training of a seemingly endless stream of digital platforms and new social media services, from Vine (now defunct) and Pinterest to Snapchat and Tumblr.

According to Trendle Polus:

> I don't know if I need to teach everyone to make a 90 second Snapchat, but I do need to teach them how to do good stories that they can put on these platforms. There is now so much out there—new channels all the time—and we can't bring all of them into the classroom or to the TV station, probably. But if they have the basics, then they can take that and go off in different directions. (Personal correspondence, April 2015)

MEANINGFUL NEWS

Students using the SMACC technology have faced other pitfalls. With a show dedicated to exploring current topics, there was a danger that such an approach, magnified by the immediacy of social media, would lead to a focus on the trivial rather than on more meaningful stories that were *not* trending on social media. However, a conscious decision was made to avoid this, by covering topics across a wide range of weighty subjects, from alternative history month and net neutrality to the slave trade. Moreover, one edition of the *Bottom Line* specifically spotlighted important news stories that had dropped from the headlines, and questioned the journalistic tendency to latch on to novelty over depth. As the students commented:

> Since we can do whatever we want basically with the show, we should do something that actually matters, that resonates with people. And college is the time to experiment so now is our best opportunity to investigate things like the slave trade or alternate Black history. I have also suggested that we do a show about taboo topics—things that we don't normally see reports on because journalism doesn't do them; things like female gender mutilation. (Anonymous student 1, personal correspondence, April 2015)

> The "topical" aspect doesn't need to limit us. If you see. … that something is not trending, you can figure out why it's not trending—maybe it's not been told in the right way. If you believe in a topic and you think it is important and that people can take interest in it, then you figure out a new way to tell that. (Anonymous student 1, personal correspondence, April 2015)

Lack of Promotion

Other potential problems have proved more intractable. A common complaint from researchers is that journalists use social media, most often, as a one-way medium to broadcast news updates, rather than to engage with audiences (Burzynski Bullard, 2013; Herrera-Damas & Hermida, 2014).

However, while *The Bottom Line* does include a segment that specifically seeks to invite and address feedback from viewers via social media, the lack of promotion of the show has meant that there is little in the way of an engaged audience to respond.

This is a blow for students, whose experience following social media through the SMACC has made them acutely aware of the importance of having an impact on an audience. As one of our interviewees put it, with no little chagrin:

> The trouble with *The Bottom Line* is that we don't have a concentrated effort to get viewers or to gather a larger audience. We work on the *Bottom Line* show but we don't work on marketing the show or raising awareness about our topic, or gather opinions about how we do our topics, and I think that could be further advanced. That's the only way to truly see a spike—to actually gather a larger audience. (Anonymous student 1, personal correspondence, April 2015)

The dilemma of *The Bottom Line* is indicative of a larger problem at the TV station, where little, if anything, is done to promote its work, whether it be to the surrounding campus or beyond. While its connection with former students and perspective employers is excellent, there has been little appetite to date, among those overseeing the station, for raising market awareness of either the daily shows or its more specialized counterparts.

DISCUSSION

Technical innovation has long had a profound effect on the purposes, practices and structure of the news industry; from the introduction of the printing press and the linotype machine through the emergence of photography and broadcast methods of dissemination (Barnard, 2012; Krause, 2011). Change is nothing new for journalism. While the adoption of digital platforms and social media is now playing an increasingly seismic role in the news industry (Burzynski Bullard, 2013; Deuze, 2007; Lee-Wright, 2010), arguably this is just the latest upheaval in a history of such innovation. As Deuze (2007) suggests: "[a]t different times in the history of the [journalism] profession, technology was (and still is) heralded as the bringer of all new threats and possibilities" (Deuze, p. 153, quoted in Barnard, 2012, p. 26).

Although journalism has always seemed a prime example of the potency of technological determinism, it would be a mistake to consider the effects of such technology without understanding the role of other, complex cultural and sociological influences that shape its dissemination and use. As Earl and Kimport (2011) argue:

> Technologies don't change societies or social processes through their mere existence but rather impact social processes through their mundane or innovative *uses*, and the ways in which the affordances of the technology are leveraged by those mundane or innovative uses. (p. 14)

The introduction of any technology includes a process of intense negotiation in which the new contends, stretches, modifies—and is resisted by—that which is already established. Such negotiation has been fierce in the news industry over more than a decade. For example, a study on the early adoption of online technology at three U.S. newspapers likened the process to a collision of cultures, where resistance to certain aspects of the technology significantly delayed the realization of the full potential of online journalism at these outlets (see Huxford & Duda, 2000).

At a time when professional journalism continues to struggle to find the best approach to incorporating social media (Burzynski Bullard, 2013; Greer & Yan, 2011), it is clear that the training that prospective journalists receive in higher education stands as a crucial factor in how this technology will eventually be utilized. Yet it is equally true that these educational venues are struggling with many of the same dilemmas as their professional counterparts, torn, for instance, between maintaining the standards of fundamental skills—such as grammar, interviewing techniques and accuracy—while also juggling the need to embrace a steady stream of new technologies. It is a balance made all the more difficult in a university organizational culture where productivity is coded as continuous improvement, improvement becomes innovation, and innovation is most easily answered by snatching at the latest technology coming down the pike, all potentially at the expense of the crucial, yet easily taken-for-granted basics on which journalism is founded. Indeed, it is likely that the dilemma may stretch even into economic considerations, with the pressure to include new digital platforms and software vying with the budgetary requirements of upgrading and maintaining established technology.

Such issues are uppermost in the minds of many at ISU. Comments by faculty in the School of Communication, during an online discussion about technology offer an intriguing perspective on striking the right balance among innovation, integration and consolidation:

> I like to think of it being like digging a tunnel. Tunneling forward isn't what takes up the most time and effort, what does that is the constant need to shore up the tunnel behind you to stop it collapsing. Indeed, re-shoring up the tunnel further back is also a constant concern as the vibrations from the digging and the exposing of fault lines has a ripple effect backwards from the drill-point. Put all your focus on moving forward and you run the very

> real risk of the whole thing falling in on your head. … We need to think of innovation as a process, rather than a bolt-on concept—innovation without integration and consolidation is not only useless, it's … counterproductive. While teaching our students important new abilities, we must also constantly be looking to correct for the ripple effects that technology will have on other skills across the program. (School of Communication, forum email, January 2015)

One approach to handling the steady stream of new technology is offered by TV-10 News Director Laura Trendle Polus, who, in a move that reflects the findings of studies focusing on professional media organizations (Herrera-Damas & Hermida, 2014), suggests that a good deal of reliance is placed on the individual student's own interest and ability to cultivate a specific media platform.

At the same time, an informal strategy that is beginning to be adopted in the School of Communication calls for each journalism course across the curriculum to focus on teaching a different social media platform, rather than each class struggling to incorporate the full breadth of such social media. One class, for example, may focus primarily on the use of Facebook to disseminate news, while another digs more deeply into Twitter.

Perhaps the most acute point of collision in the School of Communication between new technology and existing culture comes not from an immediate impact at the drill-bit, but rather in the reverberations. As is clear from the comments of our student interviewees, one powerful effect of incorporating social media at the TV station has been an increased awareness, on the part of the student news workers, of their audience, and of that audience's reaction to the stories being produced. Yet in this campus, student-run media setting, ratings play no part in either the financing of the station or in the evaluation of its success. Thus, promoting awareness of its output—or even its presence on campus—has seldom been an important concern, with the focus instead placed almost entirely on the quality of production and the appeal to potential employers. To address this, there are now plans to attach members of the School's chapter of the Public Relations Student Society of America to the station, charged with the on-going task of promoting its work.

CONCLUSION

The dance between sites of higher education and the news outlets they help to feed is both delicate and complex. It is telling that inclusion of the SMACC technology at TV-10 constitutes a relatively unusual step in this dance, entailing as it does the incorporation of technology significantly ahead of many professional news outlets.

As Trendle Polus explains:

> We're at the cutting edge at the moment. It's a place I've never wanted to be with our newsroom software or our cameras or editing, because I have seen a few other schools have

> done that and had problems. They get an infusion of money for whatever reason and buy the latest equipment, but then the places they are sending their new graduates out to work are behind them, by seven or eight years. The students can learn—but it's frustrating for all concerned. It's better that the two technologies are close. (Personal communication, April 2015)

However, training on the SMACC software is a case in which the student gains a range of skills that can be applied to less advanced technology.

> If you are learning on a really good system of this sort that has a lot of depth you can draw on that—you can look at something smaller and offer some real insight ... They may use this software down the road, or something like it as cheaper version emerges. It's ... good training. (Trendle Polus, personal communication, April 2015)

Certainly the successes of incorporating SMACC technology include both the rise in confidence of the students in the School of Communication who use it, and the marketability of these graduates as they move into the profession. For example, in his new position, the recent graduate we talked to said he has found social media to be a crucial source for story ideas and content, and expressed his amazement that the technology included in the SMACC was not incorporated in more universities' broadcast journalism training:

> Just based on what I've seen in my two months in the real world, it's the universities that are behind. TV stations are reporting on social media topics, whether their reporters are ready for the ins-and-outs of that or not. I was ready. I found myself explaining social media concepts to the veteran reporters on my first week of the job. (Anonymous recent graduate, personal correspondence, April 2015)

Whatever concerns there might be about getting out-of-step with their professional counterparts, TV-10 responded with admirable speed to incorporate the software into its reporting. Primed, perhaps, by a move in the journalism program to more fully address convergence technologies, the station was including the SMACC, both as a source of social media analysis and as a site from which to broadcast, into the newly-launched *Bottom Line* within months of the center being opened. And the dissemination of the technology across TV-10's programming continues apace, with two further special interest shows—dealing with entertainment and sport—now also beginning to feature the SMACC. At the same time, the introduction at the School of Communication of a second software program, Inception, is set to bring further use of social media into the classroom for journalism students.

Rising to the challenge of social media technology has largely proved successful for ISU's journalism program; adding essential new skills, promoting a deeper awareness of audiences and generating a buzz among students and employers alike. Yet the reverberations from the introduction of this and other digital platforms are still being felt and addressed. Adding new exercises in the classroom to ensure students are occasionally pried from their computer screens to add a real-world search for stories on campus; ensuring that social media is only one source of quotes used in producing TV and newspaper reports; focusing on a specific platform per course; and drafting in PR specialists to meet the increased need for promoting the TV station to its viewers, are prime examples of the ongoing process of integration and consolidation that must accompany technological innovation, if such innovation is to prove meaningful.

To quote again from the online discussion among ISU faculty: "We have to be smarter than the technology we teach. And we have to make sure our students are too" (School of Communication, forum email, January 2015).

REFERENCES

Barnard, S. R. (2012). *Twitter and the structural transformation of the journalistic field. How the growth of a new(s) medium is transforming journalism* (Doctoral dissertation). Retrieved from https://mospace.umsystem.edu/xmlui/bitstream/handle/10355/15860/research.pdf?

Baskarada, S. (2014). Qualitative case studies guidelines. *The Qualitative Report, 19*, 1–25.

Bor, S. E. (2014). Teaching social media journalism: Challenges and opportunities for future curriculum design. *Journalism & Mass Communication Educator, 69*, 243–255.

Burzynski Bullard, S. (2013). *Social media and journalism: What works best and why it matters.*

Retrieved from: http://www.aejmc.org/home/wpcontent/uploads/2013/08/Sue-Burzynski-Bullard.pdf

Castaneda, L., Murphy, S., & Hether, H. J. (2005). Teaching print, broadcast, and online journalism concurrently: A case study assessing a convergence curriculum. *Journalism & Mass Communication Educator, 60*, 47–56.

Deuze, M. (2007). *Media Work*. Cambridge: Polity Press.

Dimitrov, R. (2014). Do social media spell the end of journalism as a profession? *Global Media Journal Australian Edition, 8*, 1–16.

Earl, J., & Kimport, K. (2011). *Digitally enabled social change: Activism in the internet age*. Cambridge, MA: MIT Press.

Greer J., & Yan, Y. (2011). Newspapers connect with readers through multiple digital tools. *Newspaper Research Journal, 32*(4), 83–86.

Herrera-Damas, S., & Hermida, A. (2014) Tweeting but not talking: The missing element in talk radio's institutional use of Twitter. *Journal of Broadcasting & Electronic Media, 58*(4), 481–500.

Huxford, J., & Duda, N. (2000, June). *Cultures in collision: Newspapers and the internet.* Paper presented at the meeting of the International Communication Association, Acapulco, Mexico.

Kolodzy, J., Grant, A. E., DeMars, T. R., & Wilkinson, J. S. (2014). The convergence years. *Journalism & Mass Communication Educator, 69*, 197–205.

Kraeplin, C., & Batsell, J. (2011, October). *Converging with the former audience: Cross-platform partnerships slip as newsrooms focus on collaborating with the public.* Paper presented at the Convergence and Society: Journalism, Sustainability, and Media Regeneration Conference, Columbia, SC.

Kraeplin, C., & Criado, C. A. (2005). Building a case for convergence journalism curriculum. *Journalism & Mass Communication Educator, 60*, 47–56.

Krause, M. (2011). Reporting and the transformations of the journalistic field: US news media, 1890–2000. *Media, Culture & Society, 33*, 89–104.

Lee-Wright, P. (2010). Culture shock: New media and organizational change in the BBC. In N. Fenton (Ed.), *New media, old news: Journalism and democracy in the digital age* (pp. 71–86). Thousand Oaks, CA: Sage Publications.

Overholser, G. (2009). *What is journalism's place in social media?* Retrieved from http://niemanreports.org/articles/what-is-journalisms-place-in-social-media

Owen, W. F. (1984). Interpretive themes in relational communication. *Quarterly Journal of Speech, 70*(3), 274–287.

Picard, R. G. (2009). *Blogs, tweets, social media, and the news business.* Retrieved from http://niemanreports.org/articles/blogs-tweets-social-media-and-the-news-business

Skoler, M. (2009). *Why the news media became irrelevant—and how social media can help.* Retrieved from http://niemanreports.org/articles

Stake, R. (2000). Case studies. In N. Denzin & Y. Lincoln (Eds.), *Handbook of qualitative research* (2nd ed., pp. 435–454). Thousand Oaks, CA: Sage Publications.

Stassen, W. (2010). Your news in 140 characters: exploring the role of social media in journalism. *Global Media Journal African Edition, 4*, 1–16.

Vanwynsberghe, H., & Verdegem, P. (2013). Integrating social media in education. *CLCWeb: Comparative Literature and Culture, 15*(3), 1–10.

SECTION FOUR

Contextual Issues

CHAPTER THIRTEEN

Campus Media Advisors

Thoughts for the Future and Resources Available

MORTIMER W. GAMBLE AND PATRICK J. SUTHERLAND

There is evidence that college and university media advisor roles are increasingly being recognized as "professional." The College Media Association (CMA), formerly College Media Advisers, has been publishing survey results of matters related to "college and university press" regularly since the early 1980s. Two-thirds of college media advisors according to Kopenhaver (2014) now have written job descriptions compared to fewer than half in the 1980s.

Terracina-Hartman and Nulph (2012) did a study researching the credentials of instructors and media advisors. They noted "… data on leadership have the potential to ultimately alter administration definitions, classification, and evaluation of the advising position" (p. 24).

This chapter posits that, with some exceptions, media advisors who are open to new pedagogy and training, including any regular in-house training that may be available, are helping students to learn. In an increasingly assessment-based higher education environment the authors of chapter, who are longtime campus media advisors themselves, believe as well that both inexperienced and veteran college media advisors will benefit from examining their campus media's mission statements (and related support documents).

Advisors need to assertively remind faculty colleagues, students and administrators that advisors are professional educators who strive to teach, mentor, and coach within a swirling whirlwind of technological and structural change.

College media advisors and their students can build upon their professionalism and training, further develop their skill sets, and engage in scholarship via active participation in, and membership with, a variety of national membership groups, among other development efforts (Terracina-Hartman & Nulph, 2012; Tolstedt, 1994).

Campus media advisors are encouraged to become (or remain) actively involved with these national membership groups and educational foundations. (A listing of some of these organizations can be found toward the end of this chapter.) Gregory Weston, the 2014 President of College Broadcasters, Inc. (CBI) stated that support and resource organizations like CBI appear to be serving faculty media advisors better than ever before with their services (Weston, personal communication, September 8, 2014).

Blanchard and Christ (1993), as cited in Christ and Hynes (2006), wrote that explicit mission statements merit being the focus of discussions of curricula. Shouldn't college and university mission statements also help to provide guidance, inspiration, and much-needed support to extracurricular (or co-curricular) activities such as involvement in campus media? Kanigel (2012) wrote that with student newspapers, for example, mission statements provide student staffers with an opportunity to say what the campus medium's mission should be.

Campus media advisors can guide the mission-statement writing process by aligning their campus media mission statements, constitutions, and policy books, as applicable, with their academic departments' missions. The authors recommend a close examination of campus media mission statements to keep current with target audiences and values of the host department. Kanigel (2012) suggested the mission statement include answers to questions such as "Who do you want to serve? What values do you want to uphold?" (p. 200).

SHIFTS IN CAMPUS MEDIA ADVISORS' ROLES AND FUNCTIONS

Because the digital revolution has had obvious and well-publicized impacts on traditional print media, notably newspapers, the role of the campus newspaper advisor, and the roles and functions of media advisors in general, are changing as well. The campus media advisor's responsibilities have shifted in three important ways, paralleling current industry trends, practices, and needs.

One trend is to partner with students to develop content that will attract readership in the digital age of information saturation and competition for consumers' attention. A second trend is to teach strategies that give students the skills to become independent, self-employed journalists. A third trend is to link the study and practice of journalism to the wider academic offerings and teaching mission of the institution; to develop journalists who are conversant in a global "beat" consisting of the events and issues of a worldwide society.

These new opportunities and expectations for a more "assertive" campus media advisor have begun to inspire questions about the resources that may be needed for an advisor to be effective and up to date.

ASSERTIVE ADVISING: ATTRACTING AUDIENCES IN THE DIGITAL AGE

Digital technology fragments audiences. How do campus media advisors help their staffs stay relevant to audiences? Campus newspaper advising, for example, is often subject to dissenting opinions about its best practices (how hands-on or hands-off should an advisor be?) and advising continues to evolve along with the technology and media applications it employs. The customary pattern was—and still is, largely—for student newspaper staffs to labor to produce a daily or weekly broadsheet edition distributed in racks or boxes around campus. Yet, as far back as 2011, a national survey of CMA members and student newspapers found that nearly 95 percent of colleges and universities were already making their newspapers available online (Kopenhaver, 2012). The advisor's role has been just that—to advise, reminding students as needed of their journalistic, ethical, and legal responsibilities but more often than not, remaining deferential to the student team's product and avoiding the role of censor.

Structurally and functionally, some campus media have moved away from traditional advisor positions. For example, Robert Stewart, the Director of the E. W. Scripps's School of Journalism in Athens, Ohio, noted that Ohio University's newspaper *The Post* is editorially independent from the university and receives a subsidy from the University's Division of Student Affairs. *The Post* does not have an advisor *per se*, but it does have a publishing board comprised of community volunteers including one or two journalism professors. The formal power of the board primarily consists of selecting the next newspaper editor. *The Post* does have a writing coach, who usually has had substantial professional journalism experience (R. Stewart, personal communication, December 23, 2014). Digby (2006) adhered to this largely hands-off view of advising.

> The advisor's role is very limited in order to allow the students their legal rights to a free press. U.S. Courts and journalism ethicists have repeatedly said that prior review (the reading of content before it is published) is a violation of the freedom of the press. (Digby, 2006, para 2)

Digby continued,

> I never look at a story unless the students ask me to review it. And when they ask me to look at it, I need to refrain from suggesting specific changes. When asked, I will point out problems, ethical issues, sometimes even spelling errors. Sometimes, students will come to me repeatedly with questions about stories and I will see a story several times before it is published. Any reporter or editor can bring me something to review at any point during the process. When I do get a chance to review something, I try as best as possible to avoid inserting my own personal (as separate from professional) opinion about an issue. (Digby, 2006, para. 4 and para. 5)

Compared to today's news cycle, the pace of production used to be leisurely. Online production technology, however, has opened vast new opportunities for updating, illustrating, and interacting with stories; the world of instant news and immediate reaction in our society has inevitably reshaped how campus newspapers are produced and perceived. The traditional production cycle, once daily or weekly, is now hourly or updatable by the minute for campus online editions. Along with stressing the value of strong writing and newsworthiness, now the advisor's role, from a technical viewpoint, is to help teach the possibilities presented by the technology, to keep pace with what is posted, and to monitor quality and propriety of content in a format of constant, compelling change.

An even greater challenge, amid the plethora of information and images available online, is to attract and retain readership. As Reimold reminds us:

> Over the past decade digital tools and mobile platforms have rocketed journalism to a universe of innovation, interactivity and immediacy once unimaginable. Yet, without stellar content, journalism 2.0 is not worth the effort to read, watch, click on, scroll through, contribute to or connect with. Everything journalism was, is and will be rests on our ability to tell a story. And every story starts with an idea. (Reimold, 2014, para. 1)

Given the competition for the attention of information consumers, the campus newspaper advisor, if still not a censor or *de facto* manager of the medium, may nevertheless find new responsibility in working with students to develop those inspired story ideas—ideas that lead to stories of interest. On the theory that the lifespan of hard news and features has shrunk along with readers' attention spans, a campus online newspaper is ever-hungrier for good, useable content, aside from breaking stories or sidebar features that can be updated over hours or days. Working with the student editors and reporters, the advisor therefore can and should prompt discussion that will generate engaging stories. In fact, the advisor may find that taking more of an active role than ever before is necessary in production meetings by generating story ideas, in designing attractive layouts to showcase good content, and in critiquing the latest edition.

Although primary responsibility and learning opportunities still belong to the student staff, the campus newspaper, like its "real-world" counterpart, needs all the help it can muster to stay alive. The advisor has to jump in to lead, if for no other reason than to impress upon students how fragile is the relationship today between the medium and its easily distracted audience. This includes the reality that many students now are not as inclined as they once were to work within conventional print media. They need encouragement to trust and to invest in the medium of print.

Cook (as cited by Reimold, 2012) in his blog *College Media Matters* was paraphrased as advising advisors to "be a sounding board, a mentor, a cheerleader, and, to a limited extent, a friend (or at least a friendly figure)" (Reimold, 2012, para. 5).

At the same time, Cook suggests, "you need to be in the newsroom, often, but not all the time, and not standing over students' shoulders and critiquing every move while they write, design, and edit" (Reimold, 2012, para. 8).

Interim Department Chair of the Department of Journalism at Eastern Illinois University Sally Renaud indicated her program has a different type of student now. "Many still follow the lead of their teachers, mentors, and advisors. They need a lot more support, training, and encouragement" (S. Renaud, personal communication, October 3, 2014). Renaud added that college journalists may not have had the benefit of working on high-school newspapers. "Once you build the program, you have consistency. But building a program is challenging" (S. Renaud, personal communication, October 3, 2014).

ASSERTIVE ADVISING: TEACHING THE UNEMPLOYED TO BE SELF-EMPLOYED

The same technological trends that have increased the pace and production capability of traditional newsroom operations have also given rise to a growing opportunity for reporters—the self-employed journalist. Along with the downsizing of many traditional news operations, and the concurrent loss of professional news positions formerly available to journalism practitioners, have come the growth of new media platforms and a redefinition of what it can mean to be a journalist. This brand of reporter, though bound by the rules of journalistic standards and ethics, is in every sense a freelance—down to the technology he or she carries with them into the field. Where young journalists once had the structure of a newsroom and the ready mentoring of longtime editors at hand, they now often take their craft directly into the reporting arena, immersing themselves into the profession as assignment editor, copy editor, and reporter. Indeed, for many journalism schools and departments today, the independent journalist is already the career model.

Ward, writing of the increase in self-employed journalists in the United Kingdom, notes that "The evolution of the media industry has also seen journalists wearing an increasing number of editorial hats including the obligatory 'online presence' on social platforms, writing a blog, editing video footage and taking photographs." (Ward, 2011, para. 1)

To develop these skills and support the ambitions of students seeking to be independent journalists, communications programs will find the need to continue to adjust their curricular models. This will open opportunities for interdisciplinary cooperation between academic departments. For example, communications and business programs can join forces to prepare students for, in effect, operating their own businesses as independent journalists. Understanding digital production can include exposure to marketing and advertising.

The role of the college media advisor in the independent, digital age will also continue to evolve. Assertive campus media advisors can benefit from striving to stay current with academic studies related to campus media advising such as research from Boateng, Plopper, and Keith (2014). Reporting on survey data, the researchers discuss how professional mentoring and coaching of students in the area of developing good work habits might help to improve college media operations.

Such study results are shared in various forms by national support organizations at conferences, via electronic messaging boards, etc. Training and networking of advisors are vitally important in identifying and utilizing resources to benefit students, and to keep abreast of technological developments and other issues that are relevant to the independent journalist.

Organizations such as CMA and the Society for Collegiate Journalists (SCJ), which offer conferences and publications, are two well-known national groups for colleges and universities. State press associations may also offer electronic messaging boards that can offer quick feedback to advisors and students, in addition to having memberships of veteran media professionals. These individuals, though often having originated their careers in traditional newsrooms and broadcast centers, are often themselves in transition professionally, either as independent journalists, bloggers, freelance reporters, photographers, or publishers. They understand the economic realities of striking out on one's own.

A challenge for campus communications and journalism programs is simply keeping up with technological innovations, a critical issue for aspiring independent reporters and those who advise them. System and software upgrades, webcasting equipment, cameras, and, of course, laptop computers change quickly and frequently. Upper class student mentoring programs for newer student reporters can relieve some of the training pressure on advisors.

Robert Stewart, Director of the E. W. Scripps School of Journalism at Ohio University, said that campus media cultures are probably different at each institution. Stewart noted that campus media have been evolving "somewhat organically" within his school in the last decade or so. Stewart explained that while some faculty have high technical skills, most students working in campus media at Ohio University are comfortable learning new technologies on their own.

Another change Stewart sees occurring is a lessening of the need for physical space for campus media. Stewart stressed that the "virtual office" approach serves students well because that is what they'll find their employers expect them to be comfortable with. He said this entrepreneurial approach to campus media advising should be a trend for years to come (R. Stewart, personal communication, December 23, 2014).

In addition to demonstrating the possibilities of digital technology, and ensuring that independent journalists fully understand the ethics and standards required by responsible reporting, campus media advisors and their fellow faculty need to alert entrepreneurial journalists to the hazards they may encounter in the field. The tragic 2014 murders of independent journalists at the hands of the so-called ISIS terrorist group are grim reminders that on-the-ground reporting can be dangerous and deadly, and that not everyone who encounters journalists will respect the professional ethics or even the lives of reporters on the front lines of the world's wars and the suffering of those caught up in such conflicts.

A good source of ideas, perspective and inspiration can be alumni whose professional careers and experience take into account the new model of working independently and mastering multiple skills. SportsNet New York (SNY) Vice President for Marketing and Business Development Marie DeParis, a graduate of Bethany College in West Virginia, encourages students to "self-publish, write blogs and learn to write their own copy." In a 2014 interview in *Bethany Today* magazine, DeParis explained "that in an integrated media world that exists more and more online, they will very likely someday work for themselves" (Gamble, 2014, p. 21).

Sally Renaud points out that "student media alumni are some of the most loyal. They have a four-year relationship with their advisors, have traveled and eaten with them. Often (campus) bonds with student media are stronger than with any other group. I've never asked an alum to come back and been turned down" (S. Renaud, personal communication, October 3, 2014).

ASSERTIVE ADVISING: PREPARING JOURNALISTS FOR A GLOBAL BEAT

Just as they need specialized skills in reporting, editing and employing digital media, good journalists also need a broader view of the world and the issues impacting a global society. Especially if they choose to become independent, freelance journalists, students may find themselves reporting anywhere from Atlanta to Afghanistan. The amount of time that they have to get up to speed on the background of complex issues, such as economic development, regional conflicts and security, terrorism, climate change, continues to shrink. The era of instant news compels reporters to post their stories quickly, a hazard when careful news research and analysis are required.

How, therefore, does the campus media advisor teach students to report on such stories responsibly but also promptly? The answer lies in ensuring that student journalists have not only the traditional "well-rounded" education, but also that they benefit from collaborative research and reporting with students and faculty of other disciplines. Here the campus media advisor can be especially helpful in forging mutually beneficial relationships with other academic departments, accessing their curricula and resources and developing a constructive working relationship that brings additional talent and perspective to the publication. For example, not every student reporter need be a communications or journalism major; freelance or guest writers (faculty as well as students) are no less necessary in campus news reports than in off-campus ones. As the regular student reporters develop their beats and hone their skills, it can be enormously helpful for them to interact with faculty and students from other disciplines, especially for stories of ultra-current interest.

A story on Vladimir Putin's Russia, for example, would benefit from the insights and experiences of faculty and students in the areas of international relations, history, geography, languages, as well as other areas. A feature on football injuries could include interviews with coaches, players, and parents. Here again, college and university alumni can be helpful in encouraging student journalists to learn about many subjects, and to develop skill sets that are adaptable to understanding and reporting on complex issues. Says Eastern Illinois University's Sally Renaud, it's important to "find allies on campus who understand what you're doing (as an advisor) and support your students" (S. Renaud, personal communication, October 3, 2014).

At the same time, campus newspapers can highlight new developments and opportunities in various academic disciplines, compelling student reporters—and their wider audience—to learn information and gain perspective that is outside their own disciplines and interests. "What is an institution of higher education if it doesn't have a school paper that acts as a forum, challenges thinking and invites conversation?" asks Sally Renaud. "A paper gives a sense of community to a campus" (S. Renaud, personal communication, October 3, 2014).

SAMPLING OF SOME RESOURCES AVAILABLE TO MEDIA ADVISORS

This chapter provides a sampling of resources made available by national membership groups to campus media advisors. The information below is not meant to be comprehensive.

The Associated Collegiate Press (ACP) offers a critique service to its members for college newspapers and web sites. The ACP web site can be accessed at: (http://www.studentpress.org/acp/).

The Broadcast Education Association (BEA) features a Student Media Advisors Interest Division and has held campus media panel sessions and scholarly paper competitions for campus media advisors at its conventions, which are scheduled close to the dates of the National Association of Broadcasters Annual Convention in Las Vegas each spring. The BEA web site can be found at: (http://www.beaweb.org).

College Broadcasters, Inc. (CBI) has offered advisor workshops at its conventions and CBI President Greg Weston indicates regional campus media advisor workshops might be something CBI would look into if there is a demand for them. (G. Weston, personal communication, September 8, 2014). CBI also features a very active electronic messaging board. The CBI web site can be found at: (http://www.askcbi.org).

The College Media Association (CMA) has offered advisor workshops at its regular conventions. For new advisors, CMA has provided special summer workshops. CMA also offers sponsors media advisor mentoring relationships and an electronic messaging board to its members. See (http://www.collegemedia.org) for details on its digital resources.

The Intercollegiate Broadcasting System (IBS) facilitates faculty advisor roundtable discussions at its annual conventions. The IBS web site can be found at: (http://www.collegebroadcasters.us).

The National Association of Black Journalists (NABJ) works with colleges to identify and assist black students who want to become journalists. They offer a variety of scholarships and internship opportunities, plus an electronic messaging board and e-newsletter for students, organized under a student education, enrichment and development (SEED) program at: (www.nabj.org).

The National Association of Television Programming Executives (NATPE) features a foundation which offers student internships, faculty fellowships, and opportunities for faculty to work with professional media radio and television organizations for a few weeks each summer. The NATPE web site is at: (https://www.natpe.com/education).

The National Broadcasting Society-AERho (NBS-AERho) has offered advisors' breakfasts and advisors' sessions at its annual conventions. The NBS-AERho web site can be found at: (http://www.nbs-aerho.org).

The Online News Association (ONA) offers academic memberships and services including an ONA Educators' Facebook Page (closed-group for members only) that offers a discussion forum for issues related to education and hiring. Find the ONA web site at: (http://www.journalists.org).

The Poynter Institute offers a variety of web resources such as Webinars and also on-site training for faculty members and professionals such as a Backpack and Mobile Video Workshop for educators. Their web site is at: (http://www.poynter.org).

The Radio Television Digital News Association (RTDNA) features educator memberships, opportunities for student chapters, along with fellowships and scholarship opportunities. The RTDNA web site is at: (http://www.rtdna.org). The Society for Collegiate Journalists (SCJ) holds a national convention biennially which includes formal and informal meeting opportunities for campus media advisors. See the SCJ web site is at: (http://www.scj.us).

CAMPUS MEDIA ADVISORS AND THE FUTURE

The authors of this chapter have attempted to offer an informed "look ahead" at some of the challenges that will be faced by college media advisors in the years to come. What seems certain is that there will be continued changes in campus media structures, continued technology-driven training demands, time-management challenges, and pedagogy adjustments along with the time-honored enthusiasm and passion displayed by college students seeking cutting-edge jobs and careers in journalism and communications.

Author, 2014 CMA President, and professor Rachele Kanigel indicated campus media advisors should expect more emphasis on digital media including mobile platform delivery, multimedia products, and online-only newspapers (R. Kanigel, personal communication, September 23, 2014). Ohio University's Director of the E. W. Scripps School of Journalism, Robert Stewart, said that the more traditional approach of campus media advisors providing a "recipe" for operating student media (which advisors expected their campus media leaders to follow) seems to be changing to the more entrepreneurial approach of a pedagogy that encourages both creativity and development of critical thinking skills (R. Stewart, personal communication, December 23, 2014).

REFERENCES

Boateng, K., Plopper, B., & Keith, D. (2014). The habit gap: Why it may matter to student media advisers. *Adviser Update, 55*(1), 15A–16A.

Christ, W. & Hynes, T. (2006) Mission statements. In, W. Christ, Editor, *Assessing media education: A resource for educators and administrators.* Mahwah, New Jersey: Lawrence Erlbaum Associates.

Digby, D. (2006). The journalism advisor's role. Retrieved from http://niemanreports.org/articles

Gamble, M. (2014). From a small college to the big time. *Bethany Today*, Spring, 20–21.

Kanigel, R. (2012). *The student newspaper survival guide.* (2nd ed.). West Sussex: Wiley-Blackwell.

Kopenhaver, L. (2012). Still in growth mode: newspaper revenues, salaried positions grow; online editions expand as well. *College Media Review Research Annual*, 2011–2013, 49–50.

Kopenhaver, L. (2014, May 13). A journey in college media: challenges, opportunities and implications for the future. Retrieved from http://cmreview.org

Reimold, D. (2012, July 26). How to be a great student media adviser (or editor): 10 tips from a reporting and advising veteran. [Web log post]. Retrieved from http://www.collegemediamatters.com/2012/07/26/how-to-be-a-great-student-media-adviser-or-editor-10-tips-from-a-reporting-advising-veteran

Reimold, D. (2014, August 31). 1 million story ideas special: 'new hashtag phenomenon'#iam notyour stereotype. [Web log post]. Retrieved from http://www.collegemediamatters.com/2014/08/31/1-million-story-ideas-iamnotyourstereotype

Terracina-Hartman, C. M., & Nulph, R. G. (2012). *Credentialing of campus media advisors: Is there a doctor in the newsroom?* Paper presented at the meeting of the Broadcast Education Association, Las Vegas, NV.

Tolstedt, M. (1994). *Faculty advising and load relief: a survey of student radio.* Paper presented at annual meeting of the Broadcast Education Association Convention, Las Vegas, NV.

Ward, A. (2011). The career of the future? Number of self-employed journalists rise. [Web log post]. Retrieved from http://www.freelancejournalist.co.uk/blog_self_employed_freelance_journalist

CHAPTER FOURTEEN

Learning Outcomes and Student Media

GREGORY ADAMO

The organizations explored in this book would not exist if they were not a part of an educational institution. These organizations, college radio stations, college newspapers, and college television operations, are not simply located at a college or university, but are situated in places where learning is a priority. If that were not true, then we would be investigating NPR affiliates located in a number of institutions or professionally staffed television stations at major universities. At its core this book is looking at college media within the educational institution that is a college or university. As the various chapters here show, there is no one model of what a college media organization is or should be. One of the strengths of these media entities is that they approach student involvement in a variety of ways: most often as a student activity; at times as course or academically related organization; and sometimes as a lab or independent entity, usually for a limited number of student managers or editors with some of them paid.

How the educational mission of these organizations is prioritized or even thought of is often not evident. In fact, when the educational aspects of college newspapers, radio stations and other media organizations are acknowledged, they are often described as "instructional," as in instructing students in covering a news story or in being a DJ. Leadership, critical thinking, and management skills may be important to having a productive student media organization, but the relationship between building these skills through involvement in student media is rarely seen as a part of a student's education. When that acknowledgement occurs, it is usually by those in student affairs who understand their work as an important part of a college's educational mission.

Recognition of the educational outcomes of involvement in student media is most commonly found in the anecdotal stories of alumni. I have often told people that I learned as much at my college radio station as I did anywhere else during my undergraduate years at SUNY Geneseo. When I think of my education at that station, I am talking about team building, editing audio, leadership, and, of course, announcing. Beginning with an overnight DJ slot, I worked my way up to an evening show and to working in the production department. Once I obtained enough experience I was put in charge of teaching new students' basic audio production and, in the process, got my first idea that I might want to be a teacher. It was this initial undergraduate experience that led to a career running and advising college radio stations and teaching media at the college level. In that work I always tried to honor experiential learning. For me this involved things like encouraging students to learn by doing, taking ownership of their specific areas of the station, sharing their skills *with peers* and teaching new student workshops. Most importantly they needed to be willing to learn from their mistakes. While it could be thought that my experience is unique, I have found that alumni of college radio stations I have been involved with have described their experiences as an important, and sometimes central, part of their college education.

In light of calls for "evidence of student learning", experience and stories that college media alumni have recounted over the years, most likely, are not strong enough for some administrators to acknowledge as significant educational outcomes. Exploring the importance of documenting the learning that can and does take place in college media and giving examples of how that can occur is the primary goal of this chapter. But getting at "evidence" of student learning in our field can be a problem, for much of the learning that occurs in college media materializes in subtle ways and often takes place over a long period of time. I believe that the value of documenting what students learn through their involvement in college media is possible and that using student learning outcomes is one of a number of ways to do that. Using student learning outcomes can benefit students, media advisors, and student media organizations. I make this argument with the strong caveat that doing this, to the benefit of both students and the learning environment, requires that these efforts be centered with those professionals involved in student media—those of us who know the important benefits, educational and otherwise, of students working in college radio, student newspapers, and college television operations.

First, there needs to be a definition and clarification of exactly what is meant by learning outcomes. Learning outcomes and student learning are often confused or used interchangeably. Hernon (2009) helps clarify the issue in describing student outcomes as "accountability metrics that characterize institutional success in simple terms and that are important to outside stakeholders," while student learning outcomes can be seen as, "a partnership among those teaching in a program of study, and the goal of such a partnership is to improve the quality of the educational experience while holding institutions accountable to their declared mission" (p. 28). Using the latter definition shows it is not necessary to run away from the idea, as student learning outcomes can be seen as a way to involve all the educational stakeholders in helping to define what students can be or are expected to learn.

A problem in addressing student learning outcomes is that they have been closely tied to government, corporate, and foundation led educational "reform" movements. In recent years politicians and education reformers have pushed for evidence of student learning and found their answer in assessment, objectives, and learning outcomes. When the Middle States Commission on Higher Education (MSC, 2005) released its updated accreditation standards in 2002, it admitted that changes in its standard for assessment of student learning, "engendered many questions" (p. 3). In its 2005 publication "Assessing Student Learning and Institutional Effectiveness", Middle States, one of the six accreditation agencies for higher education in the United States, spelled out their expectations for assessing an institution's effectiveness. Central to their standards was, "Developing clearly articulated learning outcomes: the knowledge, skills, and competencies that students are expected to exhibit upon successful completion of a course, academic program, co-curricular program, general education requirement, or other specific set of experiences" (MSC, 2005, p. 3). This in itself should not be seen as controversial, particularly for those involved in student media as the standard includes co-curricular work.

In 2006 President George W. Bush's Commission on the Future of Higher Education, chaired by Secretary of Education Margret Spellings, published the results of their work. The commission's report included a call that "postsecondary education institutions should measure and report meaningful student learning outcomes" (Spellings, 2006, p. 23) along with endorsement of the highly controversial "value added measurements," all in a manner that for many in higher education was viewed as a top-down approach. Taubman (2009) criticized the approach:

> In an attempt to ensure that testing and stringent accountability systems would be implemented in higher education, Spellings attempted to enact new accreditation rules that would hold accreditation bodies responsible if they did not require of the college and universities testing and the specification and measurement of learning outcomes. (p. 49)

Biesta (2010) has referred to this trend as the "culture of measurement in educational policy" (p. 5).

Acknowledging how contentious this issue is, those of us working in higher education have to live with the fact that many of those that fund, administer, or accredit our colleges and universities have bought into this trend. Or possibly they have had little choice but to do so. Over the past 15 years, we have been made well aware of the push for assessing what our students learn through required workshops on formulating learning outcomes; chairs and deans who have required us to include expected outcomes in our syllabi; or by the stories of colleges cited by their accrediting organizations for not adequately addressing student outcomes. Those involved in student media may see this issue as one where only faculty teaching in the classroom need pay attention. At the same time many of us are also committed to the idea that students involved in our radio stations, newsrooms, and television studios are receiving valuable educational experiences.

One of the best ways to understand how college media can use learning outcomes may be to look to the Association of American Colleges & Universities' Essential Learning Outcomes, published in 2007 as a part of AAC&U's LEAP campaign. These outcomes are constituted in four major groups: knowledge of human cultures and the physical and natural world; intellectual and practical skills; personal and social responsibility; and integrative and applied learning. Each of these four groups has subsets giving greater detail of what the learning outcome encompass. The rubrics associated with each major group "articulate fundamental criteria for each learning outcome, with performance descriptors demonstrating progressively more sophisticated levels of attainment" (Association of American Colleges & Universities [AAC&U], 2007, Creative Thinking). Links at the end of this chapter to the full list and associated rubrics to assess the learning outcomes developed by AAC&U are included at the end of this chapter.

I see a number of areas that could and should be applied to the educational aspects of student media. For example, students are seen to gain personal and social responsibility though "civic knowledge and engagement" and "local and global intercultural knowledge and competence" (AAC&U, 2007, Creative Thinking). Looking at the rubrics through the lens of what really goes on in college media, the capstone level of the civic engagement rubric could apply to many college radio music departments and programmers. The capstone level states, "Demonstrates evidence of adjustment in own attitudes and beliefs because of working within and learning from diversity of communities and cultures. … Promotes others' engagement with diversity" (AAC&U, 2007, Creative Thinking). College radio stations' efforts to play and promote World Music immediately come to mind in meeting this standard. And I don't say this in a dry way to justify some rubric. As commercial radio became more formatted and conservative, many college radio stations broadened their playlists and were on the cutting edge of the World Music over the past 30 years. This was an important effort in broadening the knowledge and tastes of many listeners and making World Music more than a small niche on independent record labels.

A more specific example of using rubrics, may be to illustrate how a college media organization and its advisor can use them in a more comprehensive manner. The Inquiry and Analysis Value rubric, defines inquiry as, "a systematic process of exploring issues/objects/works through the collection and analysis of evidence that result in informed conclusions/judgments. Analysis is the process of breaking complex topics or issues into parts to gain a better understanding of them" (AAC&U, 2007, Inquiry and Analysis). It seems quite clear that this could be applied in a student newspaper or news web operation, where reporters have to research and evaluate information in order to form it into a news story. The specific performance elements within the rubric are: Topic selection; Existing Knowledge; Research and/or Views; Design Process; Analysis; Conclusions; and Limitations and Implications. In many ways these conform to the steps that should be followed in reporting a news story. Going from story idea, to story meetings, research, and editors' critiques shows the relationship between inquiry and analysis and what occurs daily at a student news operation.

Applying the inquiry and values rubric might encourage greater focus for the student reporters leading to better news reports. It would also give advisors a way to structure the student news work without exerting too much control, as the rubric is structured in a fairly open manner. For example, the performance element for Existing Knowledge, Research, and/or Views range on a scale from "Synthesizes in-depth information from relevant sources representing various points of view/approaches" on the high level to "Presents information from irrelevant sources representing limited points of view/approaches" on the low end of the scale (AAC&U, 2007, Inquiry and Analysis). Applying this could also encourage ties to journalism courses as instructors can analyze what goes on in the student news operation to help students not only improve their work, but to see what they do at the student paper as a part of their education. The "design process" performance element also can encourage those that might focus on the writing part of the process to think on a wider scale to see ways their story can fit into more multimedia reporting. Finally, compiling and reporting with the use of this rubric would encourage clear ties to the college's efforts to assess learning outcomes.

What may be the most applicable area for student media could be found under AAC&U's Value rubrics. This is the one for Teamwork. Defining teamwork as behaviors "under the control of individual team members," covering effort they put into team tasks, the manner of interacting with others on the team, and "the quantity and quality of contributions they make to team discussions" (AAC&U, 2007, Values). This is a value that any of us involved in college media want to develop and encourage in our students. The five criteria in the rubric are: Contributes to Team Meetings, Facilitates the Contributions of Team Members, Individual Contributions Outside of Team Meetings, and Fosters Constructive Team Climate, and Responds to Conflict. I encourage the reader to download the rubric and look at the specific levels of achievement for each criterion. Using conflict as an example, the "benchmark," or lowest level of achievement, is "passively accepts alternate viewpoints/ideas/opinions" while the "capstone" is "addresses destructive conflict directly and constructively, helping to manage/resolve it in a way that strengthens overall team cohesiveness and future effectiveness." Those involved on student board of directors, student managers, or student editorial staffs are certainly familiar with the issue of conflict. In fact, many of us can cite times when the operations of our organizations almost came to a halt over interpersonal conflict that could have been avoided if one were not so busy putting out a product every day or week. Having some type of framework to use, such as this rubric, would allow for advisors to encourage students to take a step outside the conflict issue itself to see how they might be able to address it in a constructive manner.

This brings me to the issue of how to assess for evidence of learning. In most instances assessment is done through tests, capstone projects, or student reflection papers. I am not arguing for implementing any of these, as I believe they often take away from the unique nature of experiential learning. These types of assessments become more of a chore for students or they see this work as the focus of the activity and very often something to avoid. What I am calling for is a means of assessment located at the advisor/facilitator level. Here the rubrics could be used as discussion items in classes or credit based work involving student media. For a practicum course the instructor could introduce the rubric in the first week of the semester, refer back to it in occasional class discussions, and at the end of the course have the student self-report their results on the different elements of the rubric. Students could be encouraged to articulate what they've learned, making connections between the learning as it happened and the growth students see in what they have learned by the end of the course. These self-reports would then be used as a part of the final grade. This idea could be followed in some type of self-reflection paper at the end of the course, but only in ways where they are implemented to avoid the "chore" idea I mentioned earlier. One way of doing this would be to have course discussions lead to very short written reports that, at the end of the semester, become the paper. These would form a body of evidence, a literature, that supported a certain kind of unique situation for learning. I think it critical that advisors/facilitators must be educated about these kinds of observations and reporting. Discussions among those of us in this field as to what is of value here would be helpful.

Equally important to using any of these various assessment results in students' grading, is that they should be part of any yearly reports that advisors, instructors, and student affairs professionals produce on the student media organizations they supervise. They could also be used in program reviews. The Council for the Advancement of Standards in Higher Education (CAS) Self-Assessment Guide asks for documentary evidence that includes "Research, Assessment, and Evaluation Data" (2012, p. 7).

The push for assessment, learning outcomes, and goals in recent years is something that has not necessarily pleased me. Some of the material I've received in power point presentations in college administrator run workshops, examples of things I should include in my syllabi, and material I had to read after being appointed chair of the assessment committee of my communication department, have been very structured for faculty, laying out the specifics of how we must have measureable goals for all that we teach our students. Colleges themselves are cited by accreditation boards for not having specific programs of assessment, with the threat that the college or university might lose its accreditation, the "death penalty" in higher education.

While acknowledging the debate and coming down on the side of student centered experiential learning in college media, I still believe that those of us advising or managing these organizations should look to using learning outcomes for a number of reasons. As the first part of this chapter states, this will help us better integrate into our educational institutions. With a focus on "what students are learning," we can show that we are integral to the answer. And this can be true whether we are within academic departments or student affairs, as very often it is the latter area that has to make the case whether it is important to student learning and development. Second, by situating our radio stations, student newspapers, and television operations within the efforts that prioritize student learning, we help answer the question "what is the need for this organization in this era of personal communication technology and social media?" Those of us who have seen universities sell off their FM radio licenses know the importance of having answers to this and other questions regarding our existence. Third, it will benefit educators and our students in being a bit more focused in their work. Doing this from our perspective, people who know areas of media, rather than coming top down from administrators or academicians unfamiliar with the slow and often unsteady development of students in media areas will allow us to drive and develop a process that encourages students to learn from mistakes while having fun and fostering camaraderie—things that might not be true of most classroom experiences. This work is developed and implemented by those directly situated within the organizations. Most of us are familiar with the often disastrous consequences that can come from those outside our organizations. Finally, using student learning outcomes must be seen as only one of many ways to assess the learning that goes on in student media.

As other chapters in this book show, there is a potential richness in narrative inquiry about what occurs every day in our media organizations. The last thing I would want to see is an over-emphasis on learning outcomes and assessment, as has occurred in K-12 and even in higher education in recent years. What needs to be acknowledged is that student learning outcomes, as described in this chapter, are one of many ways of understanding and assessing the education that occurs every day in student media organizations.

Resources:
Association of American Colleges & Universities VALUE Rubric Development Project.

This web site has a link to the rubrics that I refer to in this chapter. All are free, but you do have to register with the AAC&U. https://www.aacu.org/value/rubrics

REFERENCES

Association of American Colleges and Universities. (2007, January 1). *Essential learning outcomes.* Retrieved August 14, 2015, from https://www.aacu.org/leap/essential-learning-outcomes

Biesta, G. (2010). *Good education in an age of measurement: Ethics, politics, democracy.* Boulder, CO: Paradigm.

Council for the Advancement of Standards in Higher Education. (2012, August 1). *CAS self-assessment guide for student affairs units general standards.* Retrieved November 5, 2015, from http://www.cas.edu/standards

Hernon, P. (2009, March 1). *What really are student learning outcomes?* Retrieved September 23, 2015, from http://www.ala.org/acrl/sites/ala.org.acrl/files/content/conferences/confsandpreconfs/national/seattle/papers/28.pdf

Middle States Commission on Higher Education. (2005). *Assessing student learning and institutional effectiveness: Understanding middle states expectations.* Retrieved August 12, 2015, from https://www.msche.org/publications/Assessment_Expectations051222081842.pdf

Spellings, M. (2006, September). *A test of leadership: Charting the future of U.S. higher education.* Washington, DC: U.S. Department of Education, 2006. Retrieved August 14, 2015, from http://www2.ed.gov/about/bdscomm/list/hiedfuture/reports/final-report.pdf

Taubman, P. (2009). *Teaching by numbers: Deconstructing the discourse of standards and accountability in education.* New York, NY: Routledge.

CHAPTER FIFTEEN

The Teaching Hospital Model OF Journalism Education

JACQUELINE SOTEROPOULOS INCOLLINGO

The term "teaching hospital model" of journalism education has dominated discussion and debate in the academy in recent years. In 2012 a group of six foundations that offer significant financial support for journalism education issued a letter in explicit support of this metaphor and model of experientially-based learning. In a further step, the letter warned: "Schools that do not update their curriculum and upgrade their faculties to reflect the profoundly different digital age of communication will find it difficult to raise money from foundations interested in the future of news" (An open letter to America's university presidents, 2012). This endorsement of the model, and the explicit link to funding, sparked a lively philosophical debate. However, little academic research has probed the use and definitions of the term "teaching hospital model," the goals and challenges of teaching within this framework of journalism education, and its outcomes.

The research reported here analyzes how journalism educators and administrators define the term, how they apply it to their own teaching practices, whether the model replicates and/or supports traditional, legacy models of journalism, and the challenges of teaching within aspects of the model. Finally, this chapter examines criticisms of the model as hyper-professional and lacking in emphasis on emerging entrepreneurial elements of the field.

LITERATURE AND CONTEXT

The metaphor "teaching hospital model" is drawn from the practice of using medical students and residents who have graduated from medical school, in tandem with licensed doctors and specialists, to treat patients at university hospitals. In its ideal form, the medical teaching hospital integrates education, research, and patient care, in order to advance and complete physicians' training.

While use of the term "teaching hospital model" in journalism education may be recent, the concept of hands-on journalism training via content and media production is far from novel (Francisco, Lenhoff, & Schudson, 2012).

One of the earliest examples is the University of Missouri School of Journalism, which has been publishing a daily community newspaper (separate from a student-run campus newspaper) since 1908. Presently, *The Columbia Missourian* newspaper and multimedia website are managed by professional editors who are faculty members and staffed by student journalists (www.columbia missourian.com).

Other programs have implemented the idea in different ways. At Northwestern University's Medill School of Journalism, graduate students have been generating news coverage, edited by faculty members since the 1960s (http://news.medill.northwestern.edu/chicago/about/). Today, Northwestern operates newsrooms in Chicago and Washington, D.C. with student journalists' content published by media partners and on the web.

The Columbia Journalism School launched the graduate-student staffed *Columbia News Service* in the 1970s, providing news content to legacy media clients and via its website (columbianewsservice.com), and more recently launched post-graduate newsroom projects, like "The New York World" (thenewyorkworld.com), which supplies data-driven investigative content to media partners. The University of Maryland started its *Capital News Service* in 1990, with bureaus today in Annapolis, Washington, D.C. and College Park. Multimedia and broadcast content, produced by graduate students and upper-level undergraduates, is provided to regional print, broadcast and digital media clients and published on its website (cnsmaryland.org).

Since the mid-2000s, additional universities have adopted varied approaches to publishing student-produced journalism (Anderson, Glaisyer, Rothfeld, & Smith, 2011) including the University of Alabama, Arizona State University's Cronkite School of Journalism and Mass Communication, Northeastern University in Boston, and Florida International University in Miami. Another immersive program, *News21,* was launched in 2005 by the Carnegie-Knight Initiative on the Future of Journalism Education. The program, now headquartered at ASU's Cronkite School, provides free investigative content to national media organizations while participating schools cover stipends and travel costs for participants, who may be advanced undergraduates, graduate students, or recent graduates (Downie, 2011; news21.com/about-news21/).

Despite the proliferation of journalism programs that incorporate some aspect of experiential newsgathering with published content, the varied structures of these programs suggest there is no single definition of the "teaching hospital model" of journalism education.

A prominent proponent of the "teaching hospital model"—Eric Newton, then a senior advisor at the Knight Foundation—argued in 2013 that no university had fully implemented a true "teaching hospital" for journalism (Newton, 2013a). In the vision he explicated, Newton emphasized that the model would bring together students, professors, and journalism professionals under one digital roof. Students would learn by doing, and the journalism they created would benefit and engage a community. In addition to student-produced journalism, Newton suggested a true journalism "teaching hospital" would incorporate research and experimentation with new tools and techniques (Newton, 2013a, 2013b). Proponents have argued that quality student-journalism, produced in immersive experiences modeled on medical teaching hospitals, could effectively teach students, serve communities and form important partnerships with both for-profit and nonprofit media organizations (Downie, 2011).

Skeptics argued that many schools simply lack the resources to meet Newton's ideal, or that the model over-valued professionals and under-valued scholars within the academy (Basu, 2012). Academies interested in adopting new methods of training future journalists must commit to retraining faculty to teach cutting-edge skills (Reed, 2014). Others suggested that the model reinforced old practices instead of exploring new ones, thereby failing to be entrepreneurial (Mensing & Ryfe, 2013).

In a case study of a partnership between Youngstown State University and the local newspaper and public radio station via external grant funding, Francisco et al. (2012) found that the demands of producing content undermined some of the courses' larger pedagogical goals, including teaching journalism ethics and exploring contemporary media issues.

The metaphor "teaching hospital model" continues to be debated, critiqued, or praised. More institutions are implementing—or expanding—programs that incorporate elements of the model. This study examines the experiences of educators teaching in immersive "real-world" student newsrooms, bureaus or clinics.

METHODOLOGY

Semi-structured telephone interviews were conducted with ten journalism lecturers, professors, and administrators at a geographically-diverse set of American universities. Programs were identified from the literature which either actively promote the term "teaching hospital model," or prominently incorporate elements of the model into the curriculum, such as operating news bureaus, newsrooms, or published journalism projects. The researcher identified administrators and/or faculty members incorporating aspects of the model. Participants received no compensation, and before the interview they completed the online informed consent form (approved by the researcher's university *Institutional Review Board*). Administrators and faculty members were assured that neither they nor their institution would be identified, in order to allow them to speak freely without concern for promotion, tenure or grant funding. One interview subject, Eric Newton, who was with the Knight Foundation at the time of his interview and has since joined Arizona State University's Cronkite School—an outspoken proponent of the teaching hospital model—insisted that his interview be fully attributed.

Interviews transcripts were analyzed for the following broad topics and themes: definition of the term "teaching hospital model;" use of the term in talking and/or thinking about their own program and teaching; how individual programs operate; how programs serve students, audiences and legacy media organizations; challenges of operating within aspects of the model; and critiques of the "teaching hospital model" of journalism education.

ANALYSIS

Defining the "Teaching Hospital Model," Using the Term

In speeches and his digital publication *Searchlight and Sunglasses* (2013a), Eric Newton, a prominent proponent of the "teaching hospital model" of journalism education formerly with the Knight Foundation, laid out several elements that together, ideally, would form a teaching hospital of journalism education: immersion student journalism, professional mentors, teaching professors; adopting new tools and techniques; engaging a community; and research and innovation.

In an interview (May 13, 2015) however, Newton emphasized that he does not offer a "rigid recipe" or precise formula for the model: "The first observation is that we're talking about the basic idea of learning by doing. And the basic idea of immersion learning. And I think that everything else is details."

The teaching hospital metaphor, Newton said, is useful:

> For the purposes of organizing, it's good to have a name. But for the purposes of adoption, the more flexible you can be, the better off you are. And the less prescriptive you can be, the better off you are. Because the ideas and principals can be applied in almost an infinite number of ways in different places. (E. Newton, personal communication, May 2015)

Newton said, as he envisions it:

> The teaching hospital model is … learning by doing, and it's any combination of professors, professionals, students, community, innovation, and study. And so it's an aspiration, as I'm describing it. As I'm trying to describe it, it doesn't actually exist. Only some of

> the pieces exist. In some places, several pieces in combination exist. But I'm not aware of any university that has what I would describe as a fully built-out teaching hospital model. (E. Newton, personal communication, May 2015)

Under the broad, idealized definition of a "teaching hospital model" of journalism education advocated by Newton, all of the educators interviewed in this research said that they actively participate in one or more aspects of the model as part of their classroom or department curriculum. The administrators, professors and lecturers interviewed, however, held varied and diverse definitions of the term "teaching hospital model" of journalism education. The most frequently-described components included student immersion in journalism work and production of "real-world" stories and content for actual audiences—often published or broadcast via legacy media outlets.

One educator who runs a data-driven newsroom in a northeastern university said of the teaching hospital model:

> It's not just talking about it [journalism] or writing stories that only the class sees or is only posted on a class website, but actually performing the work—doing real stories that are seen by thousands of people, hundreds of people—the outside world. I think that's the best way to teach this business, because without actually publishing and getting feedback—negative or positive—on the work that you do, and that part of the process, I think it's a disservice to journalism students. (Educator A, May 2015)

Another associate professor at a mid-Atlantic university who runs a statehouse news bureau explained:

> I think it's a learn-by-doing model, in which students get experience in real-world situations. … Having a press ID if they need one for covering the capital, for example … and they have beats that they are assigned, and clients that they are serving, and expectations to produce work that will be widely distributed. … And having students … live with the consequences—of having their work read and critiqued by the public. (Educator B, May 2015)

An administrator at a southwestern university offered a more expansive definition:

> An immersive and unique learning environment that … by definition [is] not in the classroom … it's not a two-day-a-week, three hour experience. It is usually a small group of students working immersively—that could be five days a week, or two days a week, but typically full days, working with a dedicated faculty member … whose responsibilities are focused almost exclusively on that experience. [It's] learning through the creation of journalism—hopefully in an innovative way—and … engaging the community. (Educator C, March 2015)

When asked to define the "teaching hospital model" of journalism education, an associate professor at an east coast university who runs an urban multimedia newsroom said: "Um, like *Grey's Anatomy*, except less sexy? … To me, it's experiential learning, to use a previous buzz-word. … I think we've been doing it since longer than the term has been fashionable." (Educator G, April 2015). This educator—like several others interviewed—reported that he did not use the term "teaching hospital model" when speaking or thinking about classes or programs or teaching responsibilities.

A lecturer at a mid-Atlantic university who operates an immersive broadcast bureau said,

> I don't think of it as a teaching hospital. I think of it as a fully-functioning newsroom. … I choose to use the language that I'm most comfortable with. When you compare a working newsroom to a hospital, it's just semantics. I choose to deal with the reality of it—it's a working newsroom. (Educator D, March 2015)

And an associate professor at a Midwestern university said he doesn't use the term because of the limited time his student can spend at the statehouse for his elective class:

> If I had complete immersion, these students would be at the statehouse two or three days a week, really digging into the issues. Maybe [here] it's not a teaching hospital—maybe it's a 'teaching E.R.' They come in, they handle some trauma—some stories—they do some things, but they don't get the complete immersion aspect. (Educator E, March 2015)

Other educators, also working within some aspect of the model, expressed significant reservations about comparing journalism programs to medical teaching hospitals. For example, one associate professor in a Midwestern student-journalism immersion newsroom said:

> I don't know that the teaching hospital model is a great analogy for teaching journalism. … [Student] newsrooms should be more experimental … but I sincerely doubt that [at medical] teaching hospitals—I mean, it's like saying you would be doing clinical trials, without FDA approval or something, within your teaching hospital. I mean, is that really the perfect analogy? I don't know. (Educator F, May 2015)

And an assistant professor at a southern university admitted he used the metaphor, which he dislikes, only in grant applications because the term has been explicitly favored by major journalism funders. Other educators reported using the term only when discussing pedagogy with peers in the academy—not with students.

However, one administrator at a southwestern university said he intentionally uses the term to promote the model and explain the multiple ways the school's program immerses small groups of students with a dedicated professor in actual journalism production *outside* of the classroom, creating content that engages the community.

To Newton, arguments within the academy over the term are a source of frustration:

> The purpose of trying to name something is to try to bring people into the tent to look at it. But once you get in there, you see there's a vast pageant of possibilities. And so, to stand

outside the tent and criticize something you're imagining that's inside the tent—it's just an excuse for not going in. … I really think that it is better to go back to the individuals that expressed concerns and ask, what is their alternative to connecting to the whole university, for innovation, for open collaborative models, for community engagement. The attributes of the digital age—what is their way to teach that and if they have a way to do it that is effective, then terrific. But often there is no alternative, or the alternative is, "the way we do it now is fine," but it isn't. (E. Newton, personal communication, May 2015)

BENEFITS OF IMMERSIVE JOURNALISM EDUCATION

The educators interviewed—even those who do not favor the term "teaching hospital model"—were uniformly enthusiastic about the benefits to students of immersive, "real-world" journalism newsrooms, clinics or classes.

An associate professor at an immersive newsroom explained:

> I think this is a way of institutionalizing and ensuring that everybody gets hands-on experience. You know, we have so many students who have never actually been to random places interviewing people they don't know—even if they've been assigned that throughout their collegiate career. They find ways around it. There's no way around it here. You have to actually find these stories. You actually have to go out there and interview random community leaders. … And so there's no way you can get out of our journalism program without doing those things. (Educator G, March 2015)

Some found tremendous value in creating content for actual audiences. One associate professor explained:

> We have the students creating work that is not just for your teacher's eyes or for your own personal website—we reach an audience. I think that is a different experience. We have engagement with our audience. … So it's beyond just doing the work, it's also doing the work in a multimedia fashion and interacting with the community. (Educator F, April 2015)

And a lecturer who runs a broadcast newsroom explained:

> They learn to perform well under pressure, they learn to think quickly, and discern what it is they need to report on that's essential information. But at the same time, figure out how best to tell this story, because everything we do is a form of storytelling. (Educator D, March 2015)

At some institutions, the immersive experience is an elective, at others, a requirement. In many cases, the immersive experience is part of an undergraduate capstone, or taught combining undergraduate and graduate-level students before graduation. One administrator said:

> This is an experience that they're taking at the very end of their undergraduate or graduate program. They have the ability—we hope—they have the ability to then take all the things

> they've learned in their classroom up until that point, all the things they've learned in their various internships up until that point, and apply it in this immersive environment. So what we find is because of that, the learning curve is incredibly high. (Educator C, March 2015)

An associate professor in a mid-western newsroom said the breadth of story assignments helps develop journalism skills that prepare them well for careers and make them desirable hires:

> They're really forced to report all different kinds of stories, and breaking news, and in-depth stories, and produce video and shoot their own pictures and all the things that real reporters do at any decent community newspaper-slash-website, you know, it's as if they have an internship before they graduate. So they get way more hands-on experience, and that develops critical thinking abilities really fast. And there's no way that you can replicate that in a classroom … they have to make decisions more or less on their own, develop that really important ability to evaluate information, how to use it, whether to use it, how to vet it. It develops ethical muscle, because they have to make ethics decisions. (Educator F, May 2015)

This faculty member also promoted the experience as not only forcing students to be self-directed and think on their feet, but also as an advantage over others: "Like one lap on the competition, which is still pretty keen for a limited number of jobs." (Educator F, May 2015)

An educator who runs a statehouse bureau was more explicit about the professional preparation:

> I think they get out of it the fun and the realistic difficulties, the challenges, of being a journalist. They also get out of it a lot of jobs. … I think we've had good success getting students placed with clients after those clients have seen their work and kind of know the students' abilities. (Educator E, March 2015)

Most educators interviewed said the experience significantly aided students' ability to get jobs after graduation.

STUDENT JOURNALISM SERVING AUDIENCES

Many of the educators interviewed cited a responsibility to the community, and the need to fill voids in coverage left by disrupted legacy media organizations. "I think we serve a pretty good function—most of the places we cover very often don't make the news unless bad things happen," said the leader of an urban newsroom of students (Educator G, April 2015).

Other educators said they specifically have students working to fill gaps in news coverage. One explained:

> We try to do things that [audiences are] not seeing someplace else, that [because of] the cuts in newsroom budgets—there are fewer reporters doing in-depth things, and so by

> trying to fill some of that hole, we're hopefully providing people with stories they wouldn't see otherwise. (Educator A, May 2015)

An associate professor who runs a statehouse bureau said:

> We are filling the gaps that the Associated Press isn't filling. We are filling the gaps that local newspapers are not filling or don't have the opportunity to. … So we're covering some of the subsurface issues that still come to the governor's desk that wouldn't be covered. … This is what I tell the students—we are a state university. We have to provide a service to the state. We *should* be providing a service to the state. This is our way to … give the citizens of this state information that they wouldn't get, to participate in a democracy. So this is a social responsibility that happens here as well, and I think the students understand that and appreciate that. And I think that helps them holistically as a journalist—I'm not just writing stories and building a portfolio—there are people depending upon me to be in the room, to tell them what's happening, because they can't be here themselves. (Educator E, March 2015)

An educator at a public urban university explained: "I think that we should be in theory, educating the masses, not just the people that can pay to go to college" (Educator G, April 2015).

And an administrator at a different institution said:

> The community service part, I think, is really important—especially at a public university. We're doing a real service here. We're not covering, you know, the entertainment industry. We're not covering cops and robbers, and accidents on the highway. We're covering the sort of stuff that we think is both incredibly important and there's less of in legacy media. So we think it's actually serving an important community role. And you can argue—and I'm not sure how seriously universities take this—but presumably one of our charges in life is to serve our communities beyond the walls of campus. And this is a very tangible way to do that. (Educator C, March 2015)

STUDENT JOURNALISM COMPLEMENTING—OR SERVING—LEGACY MEDIA

Several educators said they specifically assign students to cover important issues and events that are unreported or underreported by legacy media organizations. But many said coverage decisions were also driven by the desire to entice their legacy media clients or partners to use students' content and bylines. While a few journalism schools immersively operate independent publications, broadcast programs or websites, many journalism schools work in partnership with a roster of legacy media clients—who either pay for student-journalists' content or receive it for free.

One academic administrator explained bluntly: "The economic model is *we* pay for it. The rationale is because of the learning, because of the experience, because of the public service. So the 'clients' in this case are all the news outlets in the region" (Educator C, March 2015).

Newton argues:

> Student scholars doing research don't weaken research. Medical students curing patients don't weaken medicine. Law students doing clinic work don't weaken law. Agricultural students in the field don't weaken agriculture. Music students playing in university orchestras don't weaken music. And so on. So no, journalism students practicing journalism don't weaken journalism. It's exactly the opposite. (E. Newton, personal communication, May 2015)

An associate professor who leads a statehouse bureau said:

> I don't see [student journalism] leading to less professional statehouse reporting—but it may lead to *different* statehouse reporting. My students will see stories that crusty veterans won't. … If anything, [our work] leads to *more* professional reporting. Veterans pick up on and learn from some of the things my students do; and my students pick up on and learn from a lot of the things the veteran reporters do … if it weren't for [students], these stories simply wouldn't be told. I sincerely doubt that the news outlets involved would hire more reporters to tell them. (Educator H, May 2015)

But one assistant professor who participates in an innovative and immersive news project worries:

> It is concerning if professionals from the news outlets do not invest in the technology of the classrooms and curricula that they rely on for content. Additionally, if professional journalists [editing student-journalists' work] are not trained in teaching and pedagogy and are not interested … in the academic and scholarly outcomes of the classroom, students are creating content that may not be any different, not any more innovative, not any more quality than the journalism that media companies have been able to cut from their ranks. (Educator J, May 2015)

CHALLENGES FOR EDUCATORS

Educators operating in immersive newsrooms or class clinics uniformly cited the difficulty of starting anew each semester with a new batch of students as one of their larger challenges. "We have to change it every 16 weeks and start over again—that's the biggest challenge. We start from scratch almost every single time," said a broadcast director (Educator D, March 2015).

A multimedia newsroom educator added:

> When the students have become good at what they do, and they're almost done a whole new group that doesn't really know very much at all is going to walk through the doors and we have to start all over. And so having any continuity and consistency in content and what we cover, that is by far the greatest challenge. (Educator B, May 2015)

Instructors who balance regular teaching loads or administrative or service expectations in addition to the immersive instruction cited additional challenges. One bureau chief who splits his time between the statehouse and the university said:

> I don't see the students enough. … They're producing content, I just don't see them. … [Sometimes] I'm in the office and they're at the statehouse. They just go, and do their work, and … sometimes I'd like a little more interaction—where I can walk [the students] through some things. Some are self-reliant. But I worry about them getting stale and not pushing themselves enough. (Educator E, March 2015)

Another educator who teaches courses and runs a statehouse bureau explained:

> I don't think [university administrators] quite understand how much time it takes to do it. I mean, this is counted just like a regular three-credit class, and to do it right—where you give students lots of interaction on every story and you raise those stories to a publishable level and you get them out to the client, that just takes a whole lot of time. … It's almost every night and every weekend … so I do wonder whether that's sustainable. I think I can pull me up and do it. I worry that when I leave, will someone else take it up? I don't know. (Educator B, May 2015)

Other educators saw difficulty remaining current in a transitioning field. One associate professor said,

> Trying out new things … we need more of that for sure. And there's a big problem here … I think faculty sometimes do not want to learn new things. And if no one pushes them to learn new things, then they're not going to be pushing forward in their newsrooms to get students to be more innovative. So, you know, people want to do what they know how to do. Some people are terrified by change. … I just think that innovation and experimentation have to be celebrated, and that's a tough thing to do, because academia tends to get pretty stodgy, I think. You know, so how do you create that kind of culture within academia? That's practically a $6 million question for journalism education. (Educator F, May 2015)

An additional challenge educators cited is the cost of launching or expanding immersive experiences. So much academy effort, said one assistant professor, is grant-driven: "My concern about these types of initiatives—that it's all crabs in a bucket, going after morsels of shit. … So how am I supposed to get to innovate, when I don't have capital to do so?" (Educator I, March 2015). In addition, he worried:

> All the profit that is gained by my students' work, by the labor that we're doing, goes directly back into [media businesses]. So the innovation and the engagement and all the sort of stuff people are talking about is around traditional aims and goals of journalism, which is to maintain a business model. And so I think that the conversation is a little disingenuous in terms of the issues of capital and the issues of labor. (Educator I, March 2015)

CRITICISMS OF THE "TEACHING HOSPITAL MODEL"

Like the assistant professor above—whose own student-journalism project has been lauded as innovative and engaging—critics of the "teaching hospital model" often fault these programs for links to old newsroom journalism structures, ties to legacy media organizations, or lack of innovation.

One associate professor who teaches in a student newsroom said she understands criticisms that immersive programs structured around traditional newsroom experiences are not innovative: "I think there's some validity to that, because it's hard to re-build the car while you're driving the car. You know, so we're producing content. We do experiment, but not often enough" (Educator F, May 2015).

One associate professor who runs an elective statehouse bureau acknowledged:

> It is the old-school way of looking at journalism education. But I'd also contend that just because it's not hugely innovative and it's not making great strides to let them be entrepreneurial; this type of experience helps them gets jobs in places where they then can experiment to be entrepreneurial. The fact of the matter is, these jobs are still available and our students still get jobs in traditional media. … I totally get that we have to be innovative and creative, and that there's so much opportunity. That's fine. But at some point, they've got to pay off their student loans, pay their rent and eat. And if this type of storytelling still allows them to do that, and it provides a public service in the tradition of journalism, why is that so wrong or so bad? Because once they get established—whatever they do, wherever they go—they still can be innovative and entrepreneurial. (Educator E, March 2015)

He added that innovation is taught elsewhere in the curriculum.

But a broadcast director said she emphasizes new tools and techniques:

> We bring into play every single tool that we can possibly imagine, and sometimes if we haven't thought it up, we tell them, "Let's see what you can come up with." I believe in forcing them to work outside the box, because it forces me to try new methods as well, and we all learn. (Educator D, March 2015)

A statehouse director said: "I don't see it as an either-or approach—I think you can teach and practice traditional journalism skills as well as push forward in a digital-first, entrepreneurial way" (Educator B, May 2015).

However, one assistant professor expressed far deeper concerns about the model:

> The consistent things I keep hearing are corporatization—they're not saying that word—but corporatization, hyper-professionalization—they're not saying that word either—skills-based learning, collaboration with the corporate model, anti-intellectualism … where the main goal is to send students into immediate practice, to benefit the corporate structure, and creating innovation that's only valuable if it's patentable. (Educator I, March 2015)

Other educators, however, said they believe immersive journalism education students could apply their skills in a variety of ways—not just in for-profit media organizations.

One associate professor—who teaches in an immersive newsroom yet has concerns about the term and the model, said:

> I think it's perhaps like anything else—if you're too focused on producing employees for a particular industry, and that industry is in a state of utter upheaval and really rapid change, you're kind of missing the mark, aren't you? So in terms of teaching reporting, I try to think more about fundamental skills that a person needs to do just lots of different things. … I think that if I can help students develop really strong critical thinking skills, really good communication skills, and teach them to think about how to use the technology … then there's lots of things that they can do. I absolutely do not think I'm going to produce newspaper reporters. What I think is that I'm producing a person who has really an extraordinary skill set for working in journalism, no matter what that looks like. (Educator F, May 2015)

CONCLUSIONS

Critiques of the "teaching hospital" metaphor in journalism education, and the teaching method of immersion in out-of-the-classroom news-gathering for actual audiences, have often been presented as a new iteration of the longstanding tension between scholars and professionals within the academy.

Newton said:

> There's a basic conflict between academia and newsrooms in this kind of situation, because newsrooms like practical, clear, impactful things. And universities thrive on understanding complex, customized, detailed kinds of things. And so it's like fresh fish and fine wine. They're just different. They're made in different kinds of ways. Nevertheless, you can still make a great meal if they are combined in a way that respects their differences. (E. Newton, personal communication, May 2015)

Newton added that criticisms of the immersive emphasis on production are merely "the new scholar-professional debate with new clothes on" (E. Newton, personal communication, May 2015).

This research sought to delve into the experiences and challenges of educators working *within* aspects of the teaching hospital model—faculty members who had doctorates, and those who came to teaching after extensive, professional journalism education.

Notably, educators' experiences and critiques are not simply a consequence of their personal academic education or professional background. Some of the passionate proponents of immersive journalism learning who were interviewed have PhDs. And some educators who perceived flaws in the term or the prescribed methods built lengthy careers in professional journalism.

These educators offered various definitions of the term "teaching hospital model," and while some embraced the metaphor, others reported that they had not adopted it or expressed concerns about its suitability.

Nonetheless, the educators interviewed—all of whom actively teach within some form of immersive student journalism beyond the classroom walls—cited benefits to students engaged in this style of learning. In addition, distributing quality student journalism to actual audiences, some educators said, was an important way for universities to serve the larger community—particularly in light of staffing and coverage cuts at legacy media organizations.

Teaching journalism within programs that aspire to be "teaching hospital models"—or smaller-scale immersive clinics—is not without its challenges, according to these educators. In addition to the systemic turnover of new cohorts of student journalists, instructors at some programs struggled to balance other teaching or administrative responsibilities, or worried about funding or sustainability. Others reported that they believed some of the criticisms of the model—built on traditional newsroom operations, links to legacy media, and difficulty in encouraging innovation—were not without some merit.

Although academic circles and trade publications have seen much debate regarding the "teaching hospital" method of journalism, more rigorous academic study of the topic—both qualitative and quantitative—is needed in order to explore the goals and learning outcomes of this method, just as has been done with medical teaching hospitals. In addition, researchers could investigate how the content produced by student-reporters reaches and serves audiences for news, and whether evidence shows that it serves to promote engagement in the public sphere.

REFERENCES

"An open letter to America's university presidents," 2012. Retrieved from http://www.knightfoundation.org/press-room/other/open-letter-americas-university-presidents/

Anderson, C. W., Glaisyer, T., Rothfeld, M., & Smith, J. (2011). *Shaping twenty-first century journalism: Leveraging a "teaching hospital model" in journalism education.* Washington, DC: New America Foundation.

Basu, K. (2012, August 10). *Not So Fast.* Retrieved from http://www.insidehighered.com/news/2012/08/10/debate-grows-over-journalism-education

Downie, L., Jr. (2011). Big journalism on campus. *AJR-American Journalism Review, 33*(3), 36.

Francisco, T., Lenhoff, A., & Schudson, M. (2012). The classroom as newsroom: Leveraging university resources for public affairs reporting. *International Journal of Communication, 6*, 21.

Mensing, D. & Ryfe, D. (2013). "Blueprint for Change: From the Teaching Hospital to the Entrepreneurial Model of Journalism Education." Retrieved from http://www.academia.edu/3412374/Blueprint_for_Change_From_the_Teaching_Hospital_to_the_Entrepreneurial_Model_of_Journalism_Education

Newton, E. (2013a). *Searchlights and sunglasses.* Retrieved from http://searchlightsandsunglasses.org

Newton, E. (2013b). *The teaching hospital—a goal for journalism education.* Retrieved from http://www.knightfoundation.org/press-room/speech/teaching-hospital-goal-journalism-education/

Reed, K. (2014). *Before the 'teaching hospital' model of journalism: 5 questions to ask.* Retrieved from http://www.niemanlab.org/2014/10/before-the-teaching-model-of-journalism-education-5-questions-to-ask/

CHAPTER SIXTEEN

Diversity, Learning AND Media Organizations

ALLAN DIBIASE AND CHARLES McKINNEY

INTRODUCTION

ALLAN DIBIASE

Public discussion about race and diversity in the USA is as prominent today as ever due in part to the extraordinary expansion of media forms and accessibility to them since the beginning of the 21st century. The acknowledged scope of diversity has widened greatly moving, beyond race and ethnicity, to rightly include disability, age, feminism and LGBT issues. The media has played a central role in framing discussions, prioritizing some groups while negating others, and making audiences aware of our increasingly diverse nation and the issues inherent therein. College media have been a part of these changes, as they do or do not reflect the increased diversity found on many college campuses and in society at large.

College media organizations, just like campuses themselves, have the potential to be places where students experience diversity for the first time, particularly as K-12 education in many parts of the United States has re-segregated. This chapter explores two main questions: How college media organizations have responded to and reflected issues of diversity in the USA and how do those involved deal with an increasing diversity within their media organizations? Simply put, what are students in media organizations learning about diversity?

The issue of race and media is nothing new in the media, and one only needs to go back to the last time race was so prominent in the national discussion to find that. Following the civil unrest of the mid-1960s President Lyndon

Johnson convened a commission called the National Advisory Commission on Civil Disorder. The 1968 Commission Report, hereafter referred to as The Kerner Report, found the media central to the way issues played out, indicting the news media for how it reported issues related to race. The Kerner Report (1988) in part focused on the lack of diversity in newsrooms:

> The absence of Negro faces and activities from the media has an effect on White audiences as well as Black. If what the White American reads in the newspapers or sees on television conditions his expectation of what is ordinary and normal in the larger society, he will neither understand nor accept the Black American. By failing to portray the Negro as a matter of routine and in the context of the total society, the news media have, we believe contributed to the Black-White schism in this country. (p. 383)

The Report provided a wide-ranging series of recommendations for reforming newsrooms in order to remedy these issues:

> Suggestions included initiatives by editors and news directors to engage in voluntary self-studies of their own news content, to develop sources within Black communities, and to assign regular beat reporting within African American neighborhoods. Additional suggestions were for newsrooms to adequately train all reporters to understand and report responsibly on events they find in those neighborhoods, to hire more Black reporters, to promote Black journalists into decision-making jobs, to engage the editors of Black newspapers in identifying stories and story perspectives, and to expand journalism education at high school and college levels for young African Americans who could enter and transform the journalism industry. (Byerly & Wilson, 2009, p. 212)

In a 2009 article, titled "Journalism as Kerner Turns 40: Its Multicultural Problems and Possibilities" published in *The Howard Journal of Communications,* Carole Byerly and Clinton Wilson II remind us of

> the Kerner Commission's own mandate that mainstream news organizations look to the leadership of African American editors and journalists for direction in doing a better job of covering the African American community. Forty years later, that concept must be extended into other multicultural corners of the society, and the leadership that their respective media provide to journalists of the future. In a nation that is now led by an African American president and the most ethnically diverse cabinet in U.S. history, the time for journalism to catch up is finally at hand. Some might argue that the profession's survival lies in its willingness to grasp the moment as its own lifeline into the future. (p. 219)

Questions regarding diversity that this chapter considers are whether and how student media organizations provide learning opportunities that lead to the development of individuals who are responsive to an increasingly diverse and pluralistic society. This is a research program of enormous scope and, preliminary to that work, we thought it productive to provide a series of reports from the field, stories that in various ways chronicle learning experiences in student media organizations across the country.

MY PERSONAL PASSION FOR STUDENT MEDIA THAT BECAME CATALYST FOR CHANGE

CHARLES McKINNEY

Student media is something that I always have been interested in since high school where I was involved in the literary magazine and student newspaper. Writing was something that inadvertently appealed to me because it was my way of expressing my unique voice, especially since I was a timid, introverted child with little to say. Somehow student media gave me the courage to speak out on issues that I supported or countered, and it gave my audience a glimpse into the world of a mysterious adolescent. I often found myself to be not only one of few males working in student media, but also one of few people of color. Thus, this situation put me in a special position to utilize the power of media to highlight people, places and events that would otherwise go unnoticed or underrepresented.

For example, when I became the editor-in-chief of my college newspaper at Immaculata University (located in suburban Philadelphia) I wanted to create and produce a special edition in honor of Black History Month. The idea was inspired by the fact that my predecessor had published a Women's History edition to celebrate Women's History Month every March. Just as she wanted to pay homage to the phenomenal women who shaped American history, so too had I desired to recognize the African American men and women who have made significant contributions to this country. People like W. E. B. DuBois, Arthur Ashe and Ella Fitzgerald were featured in this special February edition of the *Immaculatan*, the student newspaper.

My primary goal for having a Black History issue was to raise an awareness of the life and times of interesting and talented Black History figures on a predominantly White college campus. Throughout my four-year tenure on the newspaper staff, I worked mostly with White students, and whenever a new paper was published, it reflected articles, images and adverts that appealed to them, thereby mirroring what occurs in the real world of mainstream media outlets. When I assumed the leadership position of overseeing the newspaper (in my final year of matriculation) I knew changes had to be made. After all, isn't college meant to equip individuals for life in the workforce?

Because my University was becoming more culturally diverse (with a fair number of international students) I deemed it necessary to cater to the interests and experiences of all students, which explained my decision to publish a Black History issue. Also, I endeavored to recruit a more multicultural staff although this was no easy task because generally finding people to work in student media has always proved challenging. However, while locating an abundant mix of loyal staffers did not come into fruition, fortunately, there were different people willing to contribute photographs and stories to the paper (students who would not have been ordinarily inclined to do so). As a result, the newspaper depicted a broader spectrum of the student body as well as news that drew their attention.

President Obama's election in 2008 was captured in the November issue of the paper during my editorship. This historic issue brought its share of controversy on the Catholic campus I called home for four years and that had only become coed in 2005 the year I was admitted. I was adamant in my conviction to include history in the making because I believed that this event signified victory over issues of race that have plagued American society since its inception. In fact, I likened Obama's triumph to my success in becoming the first African American, male editor-in-chief of the student newspaper that same year. What an honor and privilege it was to write history in this manner. As a student media journalist I have embraced the powerful advice of social justice advocate and alternative journalist Dorothy Day (2002) who said "It is not only the journalist's duty to record history, but also to make it" (p. 74).

REPORTS FROM THE FIELD: APPROACHES TO DIVERSITY

CHARLES McKINNEY

As part of the research design for this writing, I conducted qualitative research in the form of case studies that have been solicited from student media organizations around the country. Each participant was asked to give a sense of issues concerning diversity that affect how and what students learn in these media organizations.

College of Staten Island

Gregory Adamo, Ph.D., Former General Manager

For a variety of personal and professional reasons I have had a long-term commitment to diversity. It has been important to my teaching, research, and media advising. At times, this has meant doing things at an organizational level, where issues of diversity can be integrated into management structures, leadership training, and programming philosophy. For example, the late 1980s was time when many college radio stations experienced fierce debates about what it meant to be "alternative." Some students took this so far as to adopt program formats that forbade airing music that could be considered at all commercial.

When this idea was raised at the college FM station I ran, I challenged the students to really think about what the idea of "alternative" meant, encouraging them to create a format that incorporated World Music, Rock, Jazz and other genres. This was something that would truly be alternative to what commercial radio aired. Working with the students we instituted a format that challenged the students to think creatively about alternative so that it included a good amount of music from a variety of musical genres, both American and around the World.

At a celebration for the 30th anniversary of that station, an alumnus, now a high school assistant principle, told me how much he appreciated that I was discussing and promoting diversity decades ago.

Since college radio stations, like any student organization, can tend to be a bit cliquish, I also made it a central part of my work to recruit all different types of students. I visited classes, spoke to a variety of student groups, and always emphasized in my talks at the training sessions that students must make everyone welcomed. To the student directors I would sometimes say, "Imagine if you walked into the radio station to ask about signing up and finding that everyone looked different than yourself?" I found the students were open to this approach and that these efforts were a factor in the radio station having a good number of students of color for a traditionally White college.

I would also make it known that diversity was an important issue and that if students wanted to talk with me they could. In addition, I would keep my eye out for how students of color were being treated. I believed that it could be a bit stressful for a young college student of color to be in a primarily White environment, one where students could get into heated debates about music and programming.

I remember an instance where I had to intervene, when a new student at the station, a young African American woman, was being pressured to meet the music preferences of what some of the White students defined as "alternative." For these students that meant genres like Punk/Alternative Rock. I spoke to the young woman, letting her know that I liked her show, and that Soul and Jazz were some of my favorite musical genres. In personal conversations, I simply tried to send the message that I thought she added value to the station and that if she had any problems she should come to me. Things seemed to work out, and the student worked at the radio station for a good part of her college career. We reconnected about five years ago, and, this now 40-something woman, a successful musician/composer/teacher, told me how much she appreciated what I had done for her, making her feel welcomed and appreciated, all those years ago.

Seton Hall University

Mark Maben, General Manager, WSOU 89.5 FM

WSOU is a heavy metal radio station at a Catholic university. That brings some diversity challenges right there. But in all seriousness, WSOU has made a conscious effort over the years to address the general diversity of the student staff. As our campus becomes more diverse, the station has worked to keep pace. New Jersey is one of the most diverse states in the nation, and as our signal reaches not only the most diverse counties in our state, but also the five boroughs of New York City, we want our student staff to reflect the world around us. This means racial, ethnic, gender, LGBT, religious, economic, life experience and other forms of diversity. To increase representation, we actively recruit outside the areas that might be typical targets for a college radio station. It is important to keep student organization advisers, faculty members, staff, and administrators who work with diverse or underrepresented populations in the loop about opportunities at the station.

Seton Hall is still home to many students who are the first in their families to attend college. This is especially true for the daughters and sons of immigrants, whose parents may not be able to guide them on the co-curricular activities available at college. So our station makes sure that these students, in particular, are aware of the opportunity to become involved with the station. Having students who are either immigrants or first-generation Americans *and* first-generation college students has brought a special kind of diversity to WSOU that has greatly benefited staff, adviser, and alumni. One factor that really has helped WSOU to increase staff diversity is our summer internship program for high school students. This includes working with students in Seton Hall's Upward Bound program, as well as local public, charter, and Catholic schools. We've found that exposure to WSOU as high school students not only increases their likelihood to attend Seton Hall, but also to join the station when they get to college.

While our overall staff has become more diverse, WSOU still finds it challenging to increase the gender diversity of our sports department. Each semester, we usually have only two or three female students on our sports staff, despite the fact that the station overall is almost an even 50/50 split between male and female staff members; thus, we have an active program to make people aware that we want more women on our sports staff. We will continue to address this challenge. Just like employers, if you want a diverse student media staff, you need to work at it. Since coming to Seton Hall in 2004, I have seen the radio station's staff become ever more diverse and a comfortable place for students, regardless of background, to hang out and to feel part of a tribe of student media broadcasters.

Northeastern Illinois University (NEIU)

Edie Rubinowitz, Associate Professor

Professor Rubinowitz has been teaching at NEIU for eight years in the Department of Communication, Media and Theatre. She has seen how the institution has flourished culturally with its large Latino student population. As a predominantly Hispanic-serving university, NEIU boasts a bilingual publication that highlights issues relevant to the Latino community. Hispanic Heritage Month is widely celebrated on campus of which its specific festivities are included in the publication.

In fact, students have the opportunity to take part in an investigative reporting boot camp as interns on race and poverty with *The Chicago Reporter*. This special program prepares them for real world careers in the field of investigative journalism, allowing them to learn about and to report on stories that directly impact their community as culturally different others.

Professor R, as she likes to be called, has students who intern at other media outlets that cater to co-cultural groups: “I have one [student] right now at *Extra Bilingual Newspaper*, a Spanish and English publication. I have had other students placed at *Telemundo* and *Univision's* local broadcast affiliates.” NEIU also has a cultural diversity center meant to foster an atmosphere of harmony and understanding between students of different backgrounds. In addition to Hispanic students, there is also a Polish populace as well as African Americans and Muslims from Asia, Africa and the Middle East who all comprise the University's diversity.

Loyola University

Herbert Jay Dunmore, Adjunct Professor

Mr. Dunmore is the television manager and an adjunct professor at Loyola University in Baltimore. His work encompasses freelance production and documentary filmmaking in that he shares his technical media savvy with students and staff in the academic community. He has also conducted media and software training for *The Washington Post* and *CBS News*. His passion for this field began at a young age and, fortunately, he was able to translate the hobby into a fruitful career. The television production work he does at Loyola centers on gender and race issues. “We allow students to educate each other,” says Dunmore as it relates to applying new principles and views of race and gender in student media.

When evaluating challenges associated with cultural diversity in [student] media, Dunmore pointed toward his stint at *NBC*. The initial conversation of trying to secure an interview (sound bites) can be difficult, as a person of color, unless you are really well connected. Dunmore used an example of an African American student journalist who went to North Dakota to report on a story. She was clearly a minority in a predominantly white community, but she had to know how to conduct herself professionally in order to get the job done. The way she handled the phone conversation didn't indicate her identity until she arrived on the scene. “Be the type of person that changes that perception or perspective,” Dunmore explained when representing one's self to the world. That experience served as preparation for her post-collegiate career. “It's all part of the four-year learning experience and learning how to handle those things that come at you,” says Dunmore. He notes how it's important to learn from mistakes because they simply fine-tune students' skills for life after college.

In general as it relates to media and minorities, Dunmore thinks that we need to have a better knowledge and self-understanding of where we are now. Dunmore speaks of how the student media adviser is responsible for ensuring that student journalists have a full 360 degree view of the news they report on for the student body. Having such information helps everyone to make informed decisions and to be aware of the issues from all angles. Dunmore advocates for giving back to academia by paying it forward, which is why he teaches best practices at national conferences, trains students and advisers to optimize work flow, and endorses organizations like College Broadcaster which can help students to compose stories and search for the news in broadcast media.

Students need that guidance, Dunmore asserts, so there needs to be mentorship programs and more of an understanding of how to connect with the community (instead of good PR for Black History Month). Then cultural diversity can actually become a regular part of academic life, something to be celebrated and enjoyed and not overlooked as if it were irrelevant or insignificant. Traveling and filming documentaries are other ways for people to become more media literate and culturally competent as they become enlightened to broader worldviews that challenge their traditional thinking, facilitating the expansion of their minds.

A LACK OF DIVERSITY IN STUDENT MEDIA SPARKS FRUSTRATION, DEBATES ACROSS THE COUNTRY

TARA JEFFRIES

Editor's Note: This article was originally published on November 11, 2015 in the online Student Press Law Center Newsletter http://www.splc.org/article/2015/11/lack-of-diversity-in-student-media-sparks-debates. It is reprinted here without alteration courtesy of the author and Creative Commons.

In her first newsroom, Jackie Alexander didn't see many people who looked like her. Alexander, who is black, described a culture in which she felt like a token, like an outsider, like someone whose voice wasn't heard. Sadly, Alexander said, about seven years later, not much has changed.

"It can be so demoralizing and frustrating to be in a room full of people who don't look like you and don't value your experience and your struggle," she said. "It can be really uncomfortable in a newsroom if you're the only person of color, if you identify differently. It's hard being the spokesperson for an entire race."

Alexander now serves as the media adviser at Clemson University, where she says a lack of racial diversity still pervades the pages of student publications, sometimes skewing the content in a misguided direction.

"That lack of representation in the newsroom leads to a lack of representation in the newspaper, unfortunately," she said. "The content is very flat, it's very one-sided, and our diverse students aren't reading us because they don't see themselves or their lives reflected in the newspaper. And if students attempt diverse stories, because there's no one else in the room to say, 'Hey, that looks off,' it runs the risk of stereotypical or downright offensive content."

For example, in the midst of a "See Your Stripes" campus-wide campaign encouraging students to embrace diversity at Clemson—whose mascot is an orange tiger—the *Tiger News* ran an infographic breaking down racial demographics on campus. The "white" category was depicted as orange, which Alexander said indicated to some students a bias toward whites as the "true orange" representatives on campus.

And, sometimes, it's not the stories that are published that cause problems, Alexander said. It's what student reporters don't cover. Covering issues central to diverse communities can entice students of color to read and join the newspaper, she said.

"You have to be invested in the community," Alexander said. "You have to show up, and you have to support [students of color]. You have to go to minority events, you've got to listen to their stories. As they see themselves represented, more students of color and students of diverse backgrounds will want to participate."

The pervasive whiteness of student newspapers—and the complex question of how to remedy it—has sparked dialogues across the country, particularly after a September op-ed critical of the Black Lives Matter movement in the Wesleyan University *Argus* newspaper spurred a rapid chain reaction among campus activists who complained that the newspaper lacked diverse perspectives and called for its defunding. Student government officials voted to form a group to examine the possibility of stripping the newspaper of half its funding.

The months-long debate made national news, sharply dividing journalistic onlookers trying to balance the desire for inclusiveness in publications with maintaining a free marketplace of ideas.

And in the last week, the complex relationship between the media and minority communities was thrust even further into the national spotlight, as a video of a confrontation between University of Missouri student protesters and a student photojournalist went viral. The protesters, part of the group Concerned Student 1950 that has been calling attention to racial issues on Missouri's campus, had set up a media-free safe space on the campus quad and tried to prohibit a student photographer from taking pictures.

Free speech advocates were outraged at the video. But student activists said they were trying to prevent a mostly-white media from creating "twisted insincere narratives" about their movement. The students later took down the "no media" signs, with organizers saying that the press is "important to tell our story and experiences at Mizzou to the world."

Still, some of the movement's supporters expressed frustration that the media had become the center of the story, despite Concerned Student 1950's successful campaign to get system President Tim Wolfe and university Chancellor R. Bowen Loftin to resign from their positions.

On Twitter, Alexander posted, "It's time for media to have a real conversation about WHY marginalized communities don't trust and want to speak to them. Until we examine and tackle why, interactions like this will continue to happen."

Our World Around Us Is Changing

Alexander said student media's diversity problem won't be solved simply by putting bodies in the newsroom. The culture has to change, too.

"We need to recognize that even though college is a great place for learning and developing information, for many of our students of color, it's not a very safe place," Alexander said:

> I think journalists pride themselves on being plucky and no-nonsense, and we don't subscribe to these ideas of power and privilege, but just like with the Internet and social media, we don't realize that our world around us is changing to be more inclusive. If we don't change, we're going to be left behind.

The whiteness of college newsrooms extends to the professional journalism world. Only 13 percent of journalists at daily newspapers nationwide belong to racial or ethnic minorities, according to the American Society of News Editors' *2015 Newsroom Employment Census*, a fairly stable figure compared to recent years. Twelve percent of newspapers surveyed reported that at least one of their top three editors is a minority journalist.

Clayton Gutzmore, student representative for the National Association of Black Journalists, said that poverty can create barriers to newsroom participation by students of color, who may not be able to afford to work for free or for low pay.

"NABJ is here to help the students who want to pursue journalism but don't see the funding in it immediately," he said, adding that the organization, along with others like it, offers scholarships to talented student journalists of color.

That has been one of the main demands by activists at Wesleyan, who want to use any money that is cut from the *Argus*' budget to create work-study positions at the newspaper in hopes that it will open up the newsroom to low-income students of color who can't afford to work for free.

Gutzmore also emphasized the importance of diverse newsroom leadership in student media.

"There needs to be more anchors," he said—editors and leaders that inspire minority journalists to succeed. As the former editor-in-chief of the *Observer* student newspaper at Broward College, Gutzmore said he led black staff writers by example.

"When I was in charge, I saw more black students writing," he said. "They had the courage to do that because they'd see me do it, so they'd want to emulate it. If we have diversity in our leadership, then more people can be attracted to come."

A Cycle of Apologies

At *The Daily Illini*, the campus newspaper at the University of Illinois, the newsroom staff is fairly diverse, students say. But that didn't stop the newspaper from publishing a cartoon (drawn by an outside artist) depicting a man climbing a fence on Halloween, proclaiming, "I'm going as an illegal immigrant!" The display struck many on campus as highly offensive—but not necessarily surprising. The newspaper issued an apology and held an open forum with campus diversity groups, including La Casa Cultural Latina.

Stacy Harwood, an associate professor of urban and regional planning at Illinois and co-author of a study on racial micro-aggressions on campus, said incidents like the cartoon controversy expose the growing fissure between white students and communities of color on campus.

"These things keep happening on campus, and students organize, they protest and they speak out, and I think, OK, the message was heard," Harwood said. "And then it happens again and again. I feel this great sense of disappointment."

The cartoon prompted criticism from *La Carta Nuestra*, a Latino student publication. Cristina Lucio, the publication's editor-in-chief, said the *Illini's* apology didn't address the issue of racism.

"The fact that they did not point out what was really wrong with the cartoon was really problematic," Lucio said. "It's almost as if they're apologizing that people noticed and that they got caught, rather than knowing that it's racist."

Illini staff reached out to *La Carta Nuestra* and other Latino groups on campus to discuss how to effectively cover the Latino community, but Lucio said the onus of tackling racism shouldn't be placed on students of color.

"I just personally struggle with the idea of whether people of color should be the ones to educate white folks or privileged folks on things like this," Lucio said.

Racially charged cartoons or illustrations have also caused outrage on other campuses. At SUNY Plattsburgh, student publication *Cardinal Points* drew ire on October 23 for publishing a front page article on minority students' admission rates with an illustration that offended many students with what they called stereotypical images of black poverty—a black student walking alongside boarded-up windows and a car on cinder blocks in a run-down neighborhood.

The illustration was slammed by publications such as *The Daily Beast*, which dubbed it "the most racist front page in America" and pointed out an offensive Instagram post allegedly written by the illustrator. Editors apologized for the front page, agreeing to comply with an administrative review of the publication, publish content focused on "social issues and justice" and provide staff with diversity training. *Cardinal Points* editors did not respond to the Student Press Law Center's repeated requests for comment.

The administrative involvement has worried some advocates for free speech and free press on college campuses. Campus debates about diverse media content have flared into conflicts about what is often deemed a culture of "political correctness" or "coddling minds" at universities.

In September, Louisiana State University Law Center completed a review of the current state of diversity at the law center, including its student newspaper *The Civilian*. The review noted that the *Civilian* had published comments that were considered sexist and racist and recommended the assignment of a faculty adviser.

Still, interim law school dean Bill Corbett emphasized the hands-off nature of the position.

"We don't intend to have any editorial control over it," Corbett said.

Editor-in-chief Mallory Richard noted that the magazine has covered several sensitive topics and doesn't intend to shy away from controversial territory.

"I can say emphatically that throughout the entire year thus far there has been no attempt by the administration to assert any editorial control over *The Civilian* in any way, shape or form," Richard said in a joint interview with Corbett.

The diversity task force was formed after a few racially charged incidents on campus, including an episode at an off-campus Halloween party in which a black law student said he was harassed with racial slurs by a white student.

The review also recommended that the publication adhere to "journalistic standards," which Richard said referred to standard journalistic practices of ethics and style.

Brown University's *Daily Herald* newspaper also came under fire for publishing an Oct. 5 op-ed that seemed to suggest the biological superiority of whites by conflating whiteness and agricultural advancements. The newspaper apologized for the piece, saying, "Our complacency has unfairly forced students of color to be our educators as well as our classmates."

Editors went on to say that creating a safe environment for students of color trumped free speech concerns: "People of color calling attention to racism does not constitute an overbearing power structure that will limit free speech. The oppressed by definition cannot censor their oppressor." *Daily Herald* editors and staff did not respond to the SPLC's repeated requests for comment.

Ultimately, advocates say that the point of newsroom diversity is to add more voices to the campus conversation.

"It's really important to have those connections with the communities we serve now, so that your first conversation isn't after we screwed up," Alexander said.

CONCLUSIONS

ALLAN DIBIASE

Now as familiar as the perennial warning about the eroding credibility of the press—and at times justified by it—calls for a more open and inclusive journalism come from all corners of the American newsroom, especially, though not entirely, from groups with names

> that signal their members' identity: Asian American Journalists Association, National Association of Black Journalists, National Association of Hispanic Journalists, National Lesbian and Gay Journalists Association, Native American Journalists Association. Louder and better orchestrated than ever before, the chorus of pleas for diversity in journalism includes as well the traditionally catholic voice of groups with historically broader agendas. The Society of Professional Journalists (SPJ), for example, promotes coverage of the "fullest possible range of people and issues" by encouraging journalists to seek out new and better sources (SPJ, 2002). And the American Society of Newspaper Editors (ASNE) pushes for "full and accurate coverage of our nation's diverse communities" by urging and monitoring "racial parity in newsrooms." ASNE insists that "the nation's newsrooms must reflect the racial diversity of American society by 2025 or sooner" (ASNE, 1998). (Glasser, Awad, & Kim, 2009, pp. 57–58)

The resources detailing a broad array of issues associated with student media organizations, diversity and learning indicate this as an area for further research, and, surely it is. Deeper exploration is important. But just as essential is the fact—as the stories in this chapter illustrate—there are things those of us involved in student media can do *now* to open doors and minds to a wider diversity of people. Diversity can make our organizations stronger and the stories our newspapers tell truer; the music our radio stations air and the images our television stations broadcast can become more reflective of our increasingly diverse nation. Many of the students creating this work are at a transformative age and to ignore how their work can set the stage for the rest of their lives should be acknowledged and celebrated. There is no better way to end this chapter than with the story Tah-Nehisi Coates (2015) tells in his best-selling book *Between the World and Me.* It's a story about discovering journalism near the end of his college career. Though it was not at a college newspaper, his story still illustrates the transformative nature of media work for a college student:

> All I then wanted was to write as those black people danced, with control, power, joy, warmth. I was in and out of classes at Howard. I felt that it was time to go, to declare myself a graduate of The Mecca, if not the university. I was publishing music reviews, articles, and essays in the local alternative newspaper, and this meant contact with more human beings. I had editors—more teachers—and these were the first white people I'd ever really known on any personal level. They defied my presumptions—they were afraid neither for me nor of me. Instead they saw in my unruly curiosity and softness something that was to be treasured and harnessed. And they gave me the art of journalism, a powerful technology for seekers. I reported on local D.C., and found that people would tell me things, that the same softness that once made me a target now compelled people to trust me with their stories. This was incredible, I was barely out of the fog of childhood, where questions simply died in my head. Now I could call and ask people why a popular store closed, why a show had been canceled, why there were so many churches and so few supermarkets. Journalism gave me another tool of exploration, another way of unveiling the laws that bound my body. It was beginning to come together—even if I could not yet see what the "it" was. (pp. 62–63)

Seen in this light, the questions become: What kinds and qualities of learning experiences will our media organizations provide for students? And, how will they reflect and address fundamental issues and concerns about life and living together in our increasingly diverse society?

REFERENCES

Byerly, C., & Wilson, C. (2009). Journalism as Kerner turns 40: Its multicultural problems and possibilities. *The Howard Journal of Communication*, 20, 209–221. doi: 10.1080/10646170903069920

Coates, T. (2015). *Between the world and me.* New York, NY: Spiegel and Grau.

Day, D. (2002). *Dorothy day: Writings from commonweal.* New York, NY: Liturgical.

Glasser, T. L., Awad, I., & Kim, J. W. (2009). The claims of multiculturalism and journalism's promise of diversity. *Journal of Communication*, 59(1): 57–78. doi: 10.1111/j.1460-2466.2008.01404.x

The Kerner Report. (1988). New York, NY: Pantheon Books.

Society of Professional Journalists Diversity Committee (2002). *Diversity: The whole story.* Retrieved on March 14, 2006, http://spj.org/diversity_thewhl_aug.asp

Author Biographies

Gregory Adamo, Ph.D., is Associate Professor in the School of Global Journalism and Communication at Morgan State University in Baltimore. He is the author of the book *African Americans in Television: Behind the Scenes*. Dr. Adamo's research interests include media and race, the production of culture, and entertainment television. He has taught at The College of Staten Island/CUNY, Rutgers University, and Richard Stockton College. Dr. Adamo was General Manager of WSIA-FM for more than two decades. He is an avid cyclist and hiker and is very proud of his two adult children, Kerian and Colin. His web page is www.gregory.adamo.com

Christopher J. Anderson, Ph.D. is Head of Special Collections, Archives, and Methodist Librarian for the Drew University Library in Madison, New Jersey, USA. He is also Librarian for the General Commission on Archives and History for The United Methodist Church. He is an Assistant Coach for Drew University Baseball and hosts a weekly college radio show called *The Drew Vault*.

Allan DiBiase, Ed.D. in the Social and Philosophical Foundations of Education from Rutgers University, has had a long career teaching philosophy and the philosophy of education at the City University of New York, and at Plymouth State University and Franklin Pierce University in New Hampshire. At CUNY he was Director of Student Life at the College of Staten Island. His area of specialization is the life and work of the American philosopher, John Dewey. He maintains an active career as a classical, collaborative pianist and is currently working on a book on Thoreau's *Journal*.

Laura K. Dorsey-Elson, Ph.D. and Associate Professor of Communication Studies at Morgan State University in Baltimore, Maryland. She specializes in leadership development, strategic organizational communication, diversity/global communication and conflict management. Her formal research interests include the intersection of leadership, culture and conflict in various organizational contexts. Laura is also a communication consultant and coach in the area of leadership development and shares her life with her husband, Omowale, and two children, Imani and Asante.

Leo J. Fahey is a retired Communication and Media Studies Instructor having taught at Kingsborough Community College, Jersey City State College, New York University, Elizabeth Seton College, Long Island University, Bronx Community College, Marymount Manhattan College, Molloy College and Fordham University. He has presented research at and organized panels for conferences of the Media Ecology Association, the Alternative Education Resource Organization, the International Association for Learning Alternatives and The Joint Journalism Historians. He remains an advocate for the Twice-Exceptional Child and is principal author of New York State legislation recognizing the Twice-Exceptional as a category of pupil requiring appropriate education.

Mortimer W. Gamble, Ed.D. is Vice President for Advancement at Virginia Wesleyan College, Norfolk, VA, with responsibilities for major gifts, special communications, and foundation and legislative relations. He was Director of Stewardship for The George Washington University; Vice President for Institutional Advancement, Hood College; Vice President for Institutional Advancement and Director of the Fairmont State Foundation, Inc., Fairmont State University; Vice President for Institutional Advancement, Waynesburg College; and Director of College Relations, West Virginia Wesleyan College, previously serving on the English and humanities faculty and advising the campus newspaper. His doctorate in higher education leadership is from West Virginia University.

K. Megan Hopper, Ph.D. in Communication from the University of Missouri-Columbia, is Assistant Professor of Journalism in the School of Communication, Illinois State University. She is a former print journalist and continues to engage in freelance reporting. Her research focuses on the sexual objectification of the female body in the media, the training and experiences of journalists, and the use and impact of new media. Hopper's teaching interests include introduction to mass communication, media and society, mass media theory and effects, reporting for the mass media, and media convergence. She can be reached by email at khopper@ilstu.edu

John Huxford, Ph.D. in Communication from the University of Pennsylvania, is a former journalist from Britain and currently Associate Professor of Journalism at the School of Communication, Illinois State University, Normal, IL, USA. His research interests include the visual, textual and digital construction of news, and the role of whistleblowers and anonymous sources in journalism. E-mail: jhuxfor@ilstu.edu

Jacqueline Soteropoulos Incollingo, Ph.D., is an assistant professor in the department of communication and journalism at Rider University in New Jersey. Her research interests include journalism pedagogy and user engagement with mobile news content from newspapers. She received her Ph.D. in journalism from the University of Maryland, where she also worked as the Annapolis bureau chief of the journalism school's student-powered news service. As a professional journalist, she covered politics, the legal system and crime at *The Philadelphia Inquirer* and was the Washington correspondent for *The Tampa Tribune.*

Charles McKinney holds an MA in media communications from Webster University Thailand and a BA in international business foreign language (Spanish & Chinese) with a minor in entrepreneurship from Immaculata University. Charles also has a TESOL certificate. A young adulthood career in teaching English overseas has kindled his research interests in cultural intelligence (CQ), institutional diversity and globalization. Currently serving as a TEFL Peace Corps Volunteer in Macedonia, Charles is a member of the Peace Corps Diversity and Inclusion Working Group and contributes best practices from his knowledge and experience as a global citizen. Reach him via e-mail at charles.mckinney.iv@gmail.com

Kimberly Meltzer, Ph.D., is Associate Professor of Communication at Marymount University in Arlington, VA. A former journalist, Dr. Meltzer worked for CNN, NBC, and local newspapers. She is the author of *TV News Anchors and Journalistic Tradition: How Journalists Adapt to Technology* (2010, Peter Lang). Her other work on journalism has appeared in *Journalism, Journalism Practice, The International Journal of Press/Politics, Electronic News, Encyclopedia of Journalism* and *National Civic Review*, as well as in several edited volumes. Her current work examines journalists' perspectives on commentary, opinion, and civility in news.

Len O'Kelly is Assistant Professor of Multimedia Journalism in the School of Communications at Grand Valley State University in Allendale, Michigan. In addition to teaching he also serves as the faculty advisor for the school's student radio and television stations. A product of college radio himself, he spent over 25 years in commercial radio and television before shifting to teaching full time. He earned his PhD in Media and Information Studies from Michigan State University's School of Journalism. He is also a Research Associate for the Radio Preservation Task Force, a project of the Library of Congress.

Melony Shemberger, Ed.D., is Assistant Professor of Journalism and Mass Communication at Murray State University in Murray, Kentucky. She is a former university public relations practitioner and journalist, specializing in education reporting. She continues to cover education as a freelance reporter. Shemberger has presented and published several papers and articles. A 2014 Business Journalism Professors Fellow, she is a Provost Faculty Teaching Fellow at Murray State. She is a lifetime member of the Honor Society of Phi Kappa Phi. Her doctorate in administration and supervision, with a concentration in higher education administration, is from Tennessee State University, Nashville.

Patrick J. Sutherland, Ph.D., advises WVBC Campus Radio and teaches sports communication and other subjects in the Department of Communications and Media Arts at Bethany College in West Virginia. He's authored three previous book chapters on various Communications topics, most recently a 2015 piece titled "Higher Education and the Athletics Culture," and a 2013 work titled "ESPN Transforms Sports Culture." Sutherland chaired the Communications Department at Bethany for seven-years and served one-year as Bethany's Sports Studies Coordinator in the Physical Education Department. His doctorate in Mass Communication is from Ohio University's E. W. Scripps School of Journalism. He's worked in commercial broadcasting.

Haijing Tu, Ph.D., is an assistant professor at Indiana State University. She received her Ph.D. in 2009 from the University of Massachusetts Amherst. Her research interests are in the areas of media technology, intercultural communication and social media. Her teaching efforts are currently focusing on Media and Society, Multimedia Production, and Communication Research Methods. Prior to joining ISU, she accumulated valuable teaching experience at Lynn University and Utah Valley University in the period between January 2012 and August 2013.

Leigh Landini Wright is Assistant Professor of Journalism at Murray State University. A former reporter and section editor for a Kentucky newspaper, Wright has received awards from the Kentucky Press Association, the Kentucky School Board Association and the Mid-South division of the American Cancer Society for her writing and design. Wright is a graduate of Murray State and holds an MFA in creative writing. Wright's research interests lie in the intersection of journalism practice and the classroom as well as best practices with social media in journalism and education.

Susan Kirkman Zake is an assistant professor and student media adviser in Kent State's School of Journalism and Mass Communication. She is the former Managing Editor for Multimedia & Special Projects at the *Akron Beacon Journal*, where she worked as a staff photographer and in various editing positions until 2006. Zake shares in three Pulitzer Prize team awards; for coverage of the attempted takeover of the Goodyear Tire and Rubber Co.; for A Question of Color, which examined local attitudes toward race; and for coverage of Hurricane Katrina as part of a Knight Ridder editing team.

Index

www.ingramcontent.com/pod-product-compliance
Ingram Content Group UK Ltd.
Pitfield, Milton Keynes, MK11 3LW, UK
UKHW021838100926
8556IPUK00004B/416

9 781433 124310